Formal Models of Divide-and-Conquer Queuing Systems in 5G Infrastructures

Viacheslav Kovtun

Formal Models of Divide-and-Conquer Queuing Systems in 5G Infrastructures

Springer

Viacheslav Kovtun
Vinnytsia, Ukraine

ISBN 978-3-032-06186-7 ISBN 978-3-032-06187-4 (eBook)
https://doi.org/10.1007/978-3-032-06187-4

This Springer imprint is published by the registered company Springer Nature Switzerland AG
The registered company address is: Gewerbestrasse 11, 6330 Cham, Switzerland

Preface

In recent decades, the evolution of digital technologies has led to the proliferation of highly distributed, dynamic, and heterogeneous information and communication systems (ICS). Among them, 5G-IoT infrastructures, with their inherently dense topologies, variable load patterns, and heightened vulnerability to cyber threats, have introduced new challenges in ensuring resource efficiency, operational continuity, and dependable performance. These systems increasingly underpin critical domains such as industrial automation, smart energy, transportation, and healthcare, where failures or delays are intolerable. Consequently, the demand for robust analytical models capable of capturing the complexity of such infrastructures has never been greater.

This monograph emerged from the recognition that conventional queueing theory frameworks, while effective in classical settings, often fall short when applied to multichannel systems with synchronisation mechanisms, heterogeneous service policies, and non-Poissonian traffic flows—features that are now typical of modern ICS architectures. To address this gap, the work explores and generalises a class of systems inspired by the divide-and-conquer (DC) paradigm. By formalising DC-like queueing systems, in which incoming requests are decomposed into multiple sub-requests, processed in parallel, and synchronised upon completion, the study provides a coherent mathematical foundation for analysing their behaviour under both nominal and stress conditions.

Chapters 1 and 2 establish the conceptual and mathematical core of DC-like systems, including well-known configurations such as Split-Merge (SM), Fission-Fusion (FF), Team Service Models (TSM), and Independent Server Models (ISM). These systems are examined through rigorous formulations based on two-dimensional Markov chains, exact generating functions, and the evaluation of response time and synchronisation delay. Particular attention is given to the interdependence of queues, which is often overlooked in simpler models.

The theoretical framework is further extended through the integration of Batch Markovian Arrival Processes (BMAP) and non-exponential service time distributions, including Erlang, Cox, and hyper-exponential laws. This makes it possible to characterise input traffic more realistically and to reflect bursty, correlated, and priority-driven request flows typical of real-world 5G-IoT systems. A range of advanced mathematical tools is employed, including the matrix-geometric method, order statistics, and extreme value theory, enabling the derivation of upper and lower bounds for key performance indicators.

From a methodological perspective, the monograph proposes a hybrid analytical framework that combines exact modelling with tractable approximations, enabling the evaluation of systems beyond the scope of closed-form solutions. The interplay between queueing networks, synchronisation mechanisms, and traffic dynamics is systematically explored, leading to new approximations for maximum response time, synchronisation variance, and buffer overload probabilities. Furthermore, the integration of stochastic update models, which mimic defensive behaviour under cyber threats, expands the applicability of the framework to security-aware system design.

The scientific contribution of this work lies in the formulation and generalisation of DC-like queueing systems as a unifying paradigm for modelling distributed architectures with coordinated service processes. It advances the field by bridging the gap between classical queueing models and the requirements of modern high-load heterogeneous systems, providing closed-form and asymptotic results for previously intractable configurations, demonstrating the practical use of order statistics and extreme value theory in the analysis of synchronised queueing behaviour, and laying a foundation for further work in adaptive resource management, resilience analysis, and cyber-threat-aware queueing models.

To verify the adequacy, precision, and practical applicability of the formal models introduced in Chaps. 1 and 2, Chap. 3 presents a comprehensive simulation study under variable infrastructural conditions. The analysis encompasses four key architectural configurations, including the baseline SM and ISM models, a cascaded service system with queue-based updating, and a stochastic update model reflecting cyberattack scenarios. Across a range of input flows (Poisson and BMAP), service time distributions (exponential, Erlang, Cox, and hyper-exponential), and update intensities, the study evaluates critical performance indicators such as average response time, buffer loss probability, synchronisation delay, and system recovery capability. The results empirically confirm that the proposed DC-like systems exhibit robustness under load, adaptability to traffic anomalies, and resilience against partial service disruptions. This transition from theoretical modelling to empirical validation demonstrates that the developed models can be recommended for both academic research and applied engineering tasks within high-demand 5G-IoT infrastructures.

This monograph is intended for a broad audience of specialists working at the intersection of queueing theory, distributed systems modelling, information and communications, and cybersecurity. It may serve as a theoretical reference for academic researchers, a methodological guide for systems engineers, and a basis for postgraduate education in performance modelling. The results presented here may also be of interest to designers of cloud native services, sensor networks, and next-generation intelligent infrastructure.

Vinnytsia, Ukraine Viacheslav Kovtun

Competing Interests The author has no competing interests to declare that are relevant to the content of this manuscript.

Contents

Chapter 1
IoT-5G Interconnection Modelling in «Divide and Conquer» Paradigm: A View Through Dependability

1.1 Introduction

The eternal maxim «divide et impera», like any truly ingenious imperative, finds its application not only in the relevant field, that is, state formation. This paradigm is an adequate basis for describing the processes in the field of services, military, medical, industrial systems, technologies of parallel and cloud computing, etc. In the field of information technology, the «divide and conquer» (DC) model [1–6] is the basis for formalizing algorithmic structures for recurrent decomposition of a highly complex initial problem into a finite set of elementary subtasks of the same type but smaller dimension. It is assumed that the solutions of all defined subtasks can be induced to obtain the desired solution of the original problem. Usually, the correctness of the created DC-algorithm is proved by the method of mathematical induction, and the time of its implementation either directly depends on the duration of solving the corresponding recurrent equation or is determined by the basic theorem on recurrent relations.

In the DC paradigm, we formalize the process of wireless information interaction of a set of end devices (for example, a sensor network of IoT devices) with a base station as a front-end interface of a special-purpose information system. Suppose that the base station receives a stream of incoming requests T_i, $i = \overline{1, \infty}$, each of which the front-end interface divides into K-related subrequests: $ST_1^i \overset{def}{=} ST_1$, $ST_2^i \overset{def}{=} ST_2, \ldots,$ $ST_K^i \overset{def}{=} ST_K$, which, in turn, are sent to the ordered address streams to K serving stations-servers, as shown in Fig. 1.1.

It is believed that the incoming requests have the same structure, and their decomposition is carried out taking into account the fact that each ST_k-th subrequest can be served only by a specialized k-th service station, $k = \overline{1, K}$. Processed subrequests get to the synchronization buffer, where they are until all the elements-subrequests of the corresponding input request are there. As soon as all subrequests of the corresponding input request appear in the buffer, the synchronization mechanism instantly

V. Kovtun, *Formal Models of Divide-and-Conquer Queuing Systems in 5G Infrastructures*, https://doi.org/10.1007/978-3-032-06187-4_1

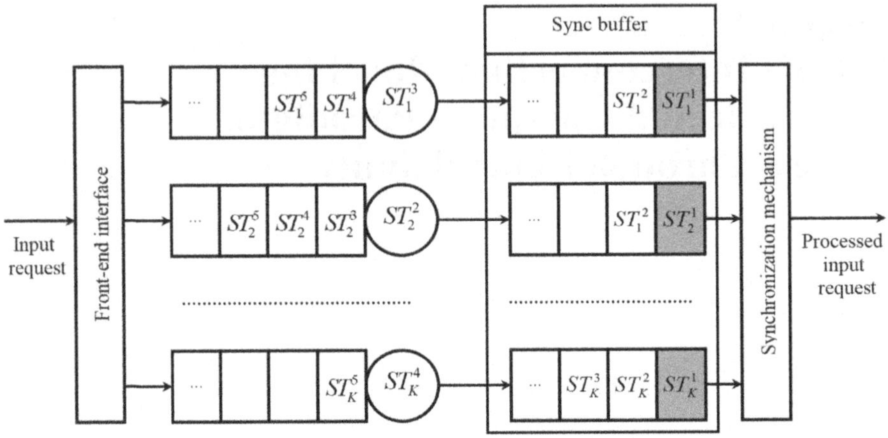

Fig. 1.1 Scheme of functioning of DC-like info-communication system

performs their induction, the result of which is the processed input request, which subsequently leaves the information system.

Representations in Fig. 1.1 process can be described in the mathematical apparatus of queuing systems, in particular, in the form of a queuing network [7–12]. In such a network, subrequests streams are processed by single-channel queuing systems with queues. Further development of this concept involves the transition to an independent analysis of these queuing systems. However, this assumption is not true because between the queues of subrequests there is a connection due to the common moments of receipt of incoming requests (the exception is when it is assumed that the system operates an unlimited number of service stations). Presented in [13, 14], the SPlitting and Matching (SPM) queueing system is a valid analogue presented in Fig. 1.1 DC-like system. That article examines the independent flow of customs control by air passengers and baggage, which are later combined at the stage of baggage delivery. This process is represented by a type $D|M|1$ queuing system, i.e. a system with a deterministic input flow and two service stations with exponential service time.

Another implementation of the DC-like queuing system is presented in Fig. 1.2. A distinctive feature of this so-called Split-Merge (SM) queuing system is that until all subrequests of the current incoming request are processed, the new incoming request is not processed by the system. Since the speed of processing subrequests by different service stations differs, the duration of processing the current input request by the SM-system is equal to the duration of processing its subrequest by the slowest service station.

Another implementation of the DC-like queuing system is presented in Fig. 1.3. A distinctive feature of this so-called Fission–Fusion (FF) queuing system is the assumption that all subrequests of all incoming requests are identical: $ST_k = ST_{k+1} = ST \forall T$, $k = \overline{1, K-1}$. Accordingly, the current incoming request may leave the information system as soon as the K subrequests are processed.

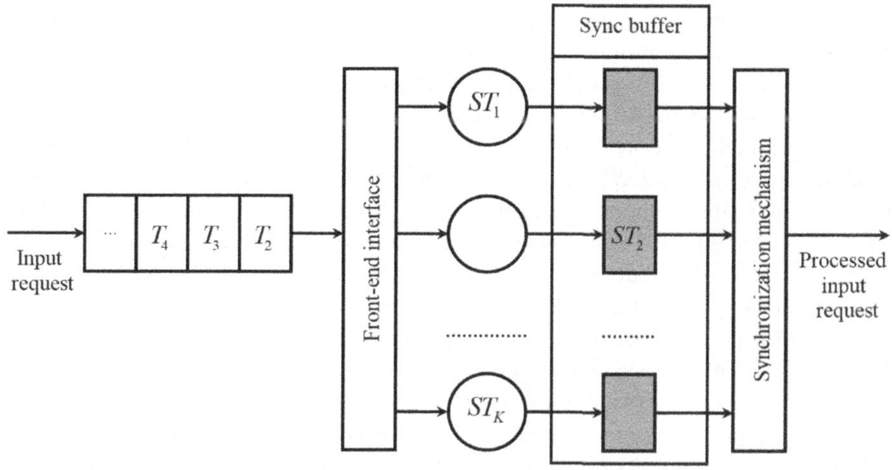

Fig. 1.2 Scheme of functioning of SM-like info-communication system

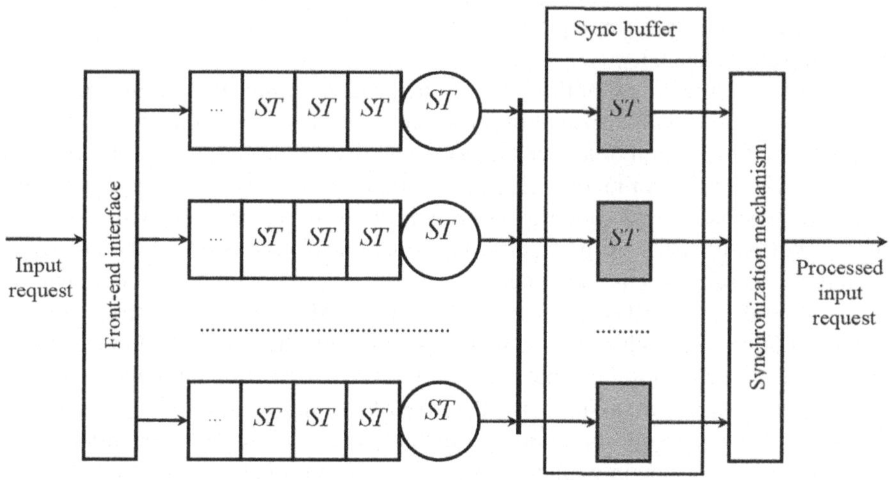

Fig. 1.3 Scheme of functioning of FF-like info-communication system

Also known is the so-called Independent Server Model (ISM), which is a variant of a multi-channel queuing system. In the ISM queuing system, $K^* \leq K$ servers are used to process the incoming request, where K^* is an integer stochastic value defined in the incoming request. Maintenance of a new input request can start only when the number of free service stations in the ISM system reaches the value K^* specified in this input request. In the study [15, 16] considered ISM-system formed by the same type of servers, the values of the service time of the corresponding input requests which are considered independent, exponentially distributed stochastic quantities. The flow of incoming requests is Poisson. A mandatory element of each input request is the stochastic value of K^*. Incoming requests are served in the order of their receipt. An incoming request can leave the information system only when all its subrequests have been processed. Since the service time for each service station is an independent value, after the subrequest service is completed, the service station becomes available for download by a new subrequest. Therefore, the service time of the input request is equal to the maximum value of the set of exponentially distributed stochastic quantities—the values of the service time of its subrequests. The authors of the article obtained expressions for the ISM system to calculate such characteristic parameters as the distribution of values of the service time of the incoming request, the distribution of the number of occupied service stations, as well as formulated the conditions for the existence of a stationary distribution.

Known is a DC-like system, formed under the condition that the service stations start and end the service of subrequests of the corresponding input request at the same time. This queuing system is known as the Team Service Model (TSM) [17–19].

Progressive is a variant of a DC-like system [20], in which the distribution of subrequests of the current input request between the service stations of the information system is carried out according to a corresponding probability matrix. This matrix is the embodiment of a system policy (strategy) for servicing incoming requests, represented by multiple subrequests, each of which is serviced by the appropriate service station independently.

However, it turned out that the study of the process of information interaction of a set of end devices with a special purpose information system in the DC paradigm is complicated by several objective factors. Even if we focus on the simplest variant when the Poisson flow of incoming requests branches into two subrequests streams, which are processed by a queuing system with two service stations with exponential distribution of service time values, the actual correlation between the processes of subrequests servicing is fundamental. Accordingly, analytically, such a queuing system should be described by a two-dimensional Markov chain with an infinite number of states for each measurement. This circumstance makes it impossible to obtain an accurate solution for the distribution of the number of subrequests in such a system. However, if we assume that the flow of incoming requests is Poisson, and the values of the subrequests servicing time are described by an exponential distribution, then we can obtain explicit expressions for marginal probabilities [21–23]. It is possible to obtain an approximate solution for the distribution of the number of subrequests for an unlimited queue of incoming requests, provided that the probability of losing the incoming request due to insufficient capacity of the queuing synchronization buffer is small enough. This task of mathematical programming can be practically solved by applying iterative methods for solving systems of linear equations of large dimensions.

Most scientists [24, 25] investigate such queuing systems in the response time metric $W_{K,\max} = \max(\xi_1, \ldots, \xi_K)$, where ξ_k is a stochastic parameter that characterizes the residence time of the subrequest ST_k in the information system before it enters the synchronization buffer, $k = \overline{1, K}$. Note that the exact expression for the average response time $W_{K,\max}$ is still obtained only for a system of type $M|M|1$ for

$K = 2$ [26]. Even for the same type of queuing systems at $K > 2$, only approximations of the value of the characteristic parameter are known. Let us investigate in more detail the known methods of determining this characteristic parameter—the response time of the system.

1.1.1 Analysis of the Queuing System of the Type M|M|*1*

In the study [27], a Markov chain model for a DC-like queuing system of the type $M|M|1$ with two service stations and an unlimited capacity synchronization buffer was continuously investigated. Subsequently, the material of this study was adapted to describe systems of type $M|G|2$ [28]. In this study, stationary and nonstationary modes of the main boundary value task (in the form of a differential equation with additional constraints) were studied using methods for research the functions of a complex variable [29]. Based on the material of the article [30], we analyse the process of obtaining and applying an exact solution for a queuing system of the type $M|M|1$ for $K = 2$, as a basic model of the studied process, presented in Fig. 1.1.

So, let in the information system with the structure presented in Fig. 1.1, implemented two service stations ($K = 2$) for processing the flow of incoming requests, each of which is divided into two subrequests. The flow of incoming requests will be considered Poisson with intensity $\lambda = 1$. Service stations of the queuing system are characterized by intensities μ_1 and μ_2. For these indicators, the inequality $1 < \mu_1 \leq \mu_2$ is performed, introduced to prevent overflow of the system synchronization buffer with processed subrequests. The operation of such a queuing system is described by a continuous Markov process $X(t) = \{x_1(t), x_2(t), t \geq 0\}$, where $x_k(t)$ is the number of subrequests of the k-th type, $k = \{1, 2\}$. Let at the time t: $x_1(t) = i$ and $x_2(t) = j$, i.e. $X(t) = \{(i, j), i, j, t \geq 0\}$.

This means that at the time t the system under study is in the state of servicing of i subrequests of the first type: $i \in ST_1$, and j subrequests of the second type: $j \in ST_2$. The probability of the studied system in this state is characterized by a stationary probability $p_{i,j}$. We formulate in terms of the generating function $P(z, w) = \sum_i \sum_j p_{i,j} z^i w^j$ [31, 32] a system of equilibrium equations for the studied system:

$$Q(z, w)P(z, w) = N(z, w), \quad |z|, |w| \leq 1, \tag{1.1}$$

where $Q(z, w) = 1 + \mu_1\mu_2 zw - \mu_1 w - \mu_2 z - z^2w^2$, $N(z, w) = \mu_2 z(w - 1)P(z, 0) + \mu_1 w(z - 1)P(0, w)$.

Assuming that $\mu = \mu_1 = \mu_2$ we describe the value of the generating function (1) at points $(z, 0)$ and $(0, w)$ by the expression

$$P(z, 0) = P(0, w) = \frac{(\mu - 1)^{\frac{3}{2}}}{\mu(\mu - z)^{\frac{1}{2}}}. \tag{1.2}$$

In [33, 34], asymptotic expressions for stationary probabilities $p_{i,0}$ and $p_{0,j}$, $i, j \to \infty$, were obtained by transforming the generating function $P(z, w)$ at the boundary points $(z, 0)$ and $(0, w)$. This fact confirms the potential for the study of the described system using a more modern mathematical apparatus.

Based on the materials of the article [35], in the article [36], the average response time $E(W_{2,\max})$ of a queuing system with two symmetric service stations is analytically described as

$$E(W_{2,\max}) = \left(H_2 - \frac{1}{8}\rho\right)E(W_{1,\max}) = \frac{12-\rho}{8}E(W_{1,\max}),$$

where $E(W_{1,\max}) = \frac{1}{\mu-\lambda}$ is the average response time of a system of type $M|M|1$, $H_2 = 1,5$; λ is the intensity of the Poisson flow of incoming requests, $\mu = \mu_1 = \mu_2$ is the symmetric intensity of service of exponential flows of subrequests on both service stations of the studied system, $\rho = \lambda/\mu$ is the load on each service station (observance of condition $\lambda < \mu$ guarantees that $\rho < 1$).

The basis for deriving $E(W_{2,\max})$ was the assumption that the required mathematical expectation of the values of the processing time of the incoming request (average response time) by a queuing system with two symmetric service stations is the sum of the average time $E(W_{1,\max})$ and the average time $E(S)$ waiting the processed first subrequest in the synchronization buffer for complete the processing of the second subrequest: $E(W_{2,\max}) = E(W_{1,\max}) + E(S)$. Little's theorem relates the parameter $E(S)$ and the average number N of subrequests that have passed the information system service station and are waiting for their pair in the synchronization buffer: $N = 2\lambda E(S)$.

Take into account the probability that the number of subrequests in the queue to the first service station exceeds the number of subrequests in the queue to the second service station by k units, by introducing the parameter q_k. Considering the symmetry of the output characteristics of the service stations of the system under study, the equality $q_k = q_{-k}$, $k \geq 1$, is satisfied. This fact allows us to write: $N = 2\sum_{k=1}^{\infty} kq_k$, $E(S) = \frac{1}{\lambda}\sum_{k=1}^{\infty} kq_k$. From the material presented in the research [37] we obtain

$$q_k = \sum_{i=k}^{\infty} p_{i,0}. \tag{1.3}$$

Taking into account (1.3), we write: $q_k = q_{l+1} + p_{k,0}$, where $p_{k,0}$ is the probability that there are k subrequests in the queue for the first service station and none in the queue for the second service station. Considering expression (1.2), we further write

$$P(z, 0) = P(0, z) \stackrel{def}{=} P(z) = \sum_{i=0}^{\infty} p_{i,0}z^i = \frac{(1-\rho)^{\frac{3}{2}}}{(1-\rho z)^{\frac{1}{2}}}. \tag{1.4}$$

Express the average waiting time for synchronization $E(S)$ taking into account expression (1.3):

$$E(S) = \frac{1}{\lambda}\sum_{k=1}^{\infty} k \sum_{i=k}^{\infty} p_{i,0} = \frac{1}{\lambda}\sum_{i=1}^{\infty} p_{i,0} \sum_{k=1}^{i} k = \frac{1}{2\lambda}\sum_{i=1}^{\infty} i(i+1)p_{i,0}. \tag{1.5}$$

Find the first two derivatives of the generating function $P(z)$ obtained in expression (1.4) at the point $z = 1$:

$$\left.\frac{dP(z)}{dz}\right|_{z=1} = \frac{1}{2}\rho = \sum_{i=1}^{\infty} i p_{i,0},$$

$$\left.\frac{d^2P(z)}{dz^2}\right|_{z=1} = \frac{3}{4}\frac{\rho^3}{1-\rho} = \sum_{i=1}^{\infty} i^2 p_{i,0} - \sum_{i=1}^{\infty} i p_{i,0},$$

and substitute the obtained analytical expressions in expression (1.5):

$$E(S) = \frac{1}{2\lambda}\left(\sum_{i=1}^{\infty} i^2 p_{i,0} - \sum_{i=1}^{\infty} i p_{i,0}\right) = \frac{1}{2\lambda}\left(\frac{\rho^2 + 2\rho}{4(1-\rho)} + \frac{1}{2}\rho\right) = \frac{\rho(4-\rho)}{8\lambda(1-\rho)}. \tag{1.6}$$

Considering expression (1.6) for the average response time of a queuing system with two symmetric service stations of type $M|M|1$, we can write

$$E\left(W_{2,\max}\right) = E\left(W_{1,\max}\right) + E(S) = \frac{12-\rho}{8}\frac{1}{\mu-\lambda} = \frac{12-\rho}{8}E\left(W_{1,\max}\right). \tag{1.7}$$

The classical approach to the approximation of functional dependence $E\left(W_{K,\max}\right)$ for a queuing system with K symmetric service stations of type $M|M|1$ is presented in studies [38, 39]. It is proposed to consider the system under study, assuming the independence of the values of the average times of stay of subrequests in it. That is, consider the system under study as a queuing network formed by independently parallel functioning queuing systems of type $M|M|1$. The authors argue the admissibility of this assumption by saying that the stochastic value of the average time of stay of subrequests in the system under study is positively associated.

As is known from classical theory [40, 41], stochastic quantities ξ and η are positively associated if the covariance of arbitrary pairs of corresponding non-descending functions $f(\xi)$ and $g(\eta)$ is nonnegative: $E(f(\xi)g(\eta)) \geq E(f(\xi))E(g(\eta))$. One of the properties of positively associated stochastic quantities, the expression

$$P\left(\max_{1\leq i\leq K} \xi_i > x\right) \leq 1 - \prod_{i=1}^{K} P(\xi_i < x)$$

is the desired upper limit for the average response time $E(W_{K,\max})$ of the studied queuing network with symmetric service stations of type $M|M|1$.

Assuming that $\rho \to 0$ we can obtain the lower limit for the average response time $E(W_{K,\max})$ of the studied system as the maximum of K independent, equally distributed stochastic quantities with average $1/\mu$:

$$H_K(1/\mu) \le E(W_{K,\max}) \le H_K E(W),$$

where $H_K = \sum_{i=1}^{k} \frac{1}{i}$ is the partial sum of the harmonic series.

The upper and lower limits of the value of the parameter $E(W_{K,\max})$ increase with the same velocity equal to H_K, and, for sufficiently large values of K, are of the order of $O(\ln K)$. Accordingly, starting from the parameter $E(W_{2,\max})$ defined by expression (1.3), we can write that $E(W_{K,\max}) \approx E(W_{2,\max})S_K(\rho)$, where $S_K(\rho)$ is the scaling factor increasing with velocity $O(\ln K)$. Therefore, the general form of approximation of the values of the parameter $E(W_{K,\max})$ is as follows: $S_K = \alpha(\rho) + \beta(\rho)H_K$. Considering that by definition: $S_2(\rho) = 1$, and, respectively, $\beta(\rho) = \frac{1-\alpha(\rho)}{H_2}$, we write

$$S_K = \alpha(\rho) + \frac{1-\alpha(\rho)}{H_2} H_K, \; K \ge 2.$$

In a study [42] as a result of simulation, it was determined that $\alpha(\rho) \approx 4\rho/11$. Summarizing the above, we formalize the average response time $E(W_{K,\max})$ of the queuing network with K symmetric service stations of type $M|M|1$ as

$$E(W_{K,\max}) \approx \left(\frac{H_k}{H_2} + \frac{4}{11}\left(1 - \frac{H_K}{H_2}\right)\rho\right)\frac{12-\rho}{8}\frac{1}{\mu-\lambda}, \; K \ge 2. \tag{1.8}$$

In the manuscript [43], the authors empirically estimated the error of approximating the values of the parameter $E(W_{K,\max})$ using expression (1.8). For the studied range of values: $2 \le K \le 32$, the approximation error did not exceed 5%.

Regardless of the described, a similar queuing network with K symmetric service stations of type $M|M|1$ was investigated in [44, 45]. In this paper, the desired approximation of the values of the parameter $E(W_{K,\max})$ was defined as the arithmetic mean of the upper and lower limit values of this parameter. If the upper limit value of the parameter $E(W_{K,\max})$ was determined similarly to the above, the lower limit value was determined by analysing the average response time of an equivalent queuing system with the non-parallel organization of the service station [46, 47]:

$$\frac{1}{\mu}\left(H_K + \rho\sum_{k=1}^{K}\frac{1}{k(k-\rho)}\right) \le E(W_{K,\max}) \le \frac{H_k}{\mu}\left(1 + \frac{\rho}{1-\rho}\right)$$
$$= \frac{1}{\mu}\left(H_K + \rho\sum_{k=1}^{K}\frac{1}{k(1-\rho)}\right).$$

Next, the arithmetic mean of the found limit values of the parameter $E\left(W_{K,\max}\right)$ is calculated:

$$E\left(W_{K,\max}\right) \approx \frac{1}{\mu}\left(H_K + \frac{\rho}{2(1-\rho)}\left(\sum_{k=1}^{K}\frac{1}{k-\rho} + (1-2\rho)\sum_{k=1}^{K}\frac{1}{k(k-\rho)}\right)\right). \tag{1.9}$$

In [48, 49] another concept of regression analysis of the response time of DC-like systems is proposed. This concept is known as Heavy and Light Traffic Interpolation Approximations (H&LTIA). This concept surpasses the methods generalized by expressions (1.8) or (1.9) in that it potentially allows to analyse the process of functioning of systems not only with the Poisson distribution of intensity of the flow of input requests λ and an exponential distribution of time of servicing of corresponding subrequests μ. If the studied system operates in low load mode (the intensity of the flow of input requests λ was small compared to the value of the intensity of the flow of subrequests μ: $\lambda \ll \mu$), then the desired asymptotic analytical dependences for characteristic parameters (in particular, for system response time) can be obtained by considering the latter as appropriate functions distribution from λ. The decomposition of the obtained distribution functions into the Taylor series in the form of polynomials from λ to n is the desired analytical dependences.

Only polynomials of zero and first degrees find practical application. If the studied system operates in high load mode (the flow of input requests λ approaches the value of the intensity of the flow of subrequests μ: $\lambda \approx \mu$), then the interpolation of the desired analytical dependence for the characteristic parameter—the response time of the system, is carried out using the key parameter $\beta \in [0, 1]$. At $\beta = 0$ the flow of incoming requests is considered to be deterministic and at $\beta = 1$ the time of servicing of subrequests by service stations of the investigated DC-like system is considered deterministic, each of which is interpreted as a queuing system of $D|GI|1$ and $GI|D|1$ types.

Adaptation of the H&LTIA method for analysis of a queuing network with K symmetric service stations of type $M|M|1$ allows to formulate an expression for approximation of values of the response time of such system:

$$E\left(W_{K,\max}\right) \approx \frac{1}{\mu-\lambda}\left(H_K + \frac{\lambda}{\mu}(V_K - H_K)\right),\ K \geq 2,\ 0 \leq \lambda \leq \mu, \tag{1.10}$$

where

$$V_K = \sum_{i=1}^{K}\binom{K}{i}(-1)^{i-1}\sum_{m=1}^{i}\binom{i}{m}\frac{(m-1)!}{i^{m+1}}.$$

If the values of the characteristic parameters λ and μ are described, respectively, by the Erlang distribution of the second order: $A(x) = 1 - (1 + 2\lambda x)e^{-2\lambda x}$, $B(x) = 1 - (1 + 2\mu x)e^{-2\mu x}$, $x \geq 0$, then expression (1.10) takes the form

$$E\left(W_{K,\max}\right) \approx \frac{1}{\mu - \lambda}\left(F_K + \frac{\lambda}{\mu}\left(\frac{1}{2}V_K - F_K\right)\right),\ K \geq 2,\ 0 \leq \lambda \leq \mu, \tag{1.11}$$

where

$$F_K = \frac{1}{\mu}\sum_{r=1}^{K}\binom{K}{r}(-1)^{r-1}\sum_{m=1}^{r}\binom{r}{m}\frac{m!}{2r^{m+1}}.$$

Mixed cases were also investigated in [50, 51]. In particular, if, respectively, the value of the characteristic parameter λ is described by a second-order Erlang distribution and the value of the characteristic parameter μ is described by an exponential distribution: $A(x) = 1 - (1 + 2\lambda x)e^{-2\lambda x}$, $B(x) = 1 - e^{-\mu x}$, $x \geq 0$, then expression (1.10) takes the form

$$E\left(W_{K,\max}\right) \approx \frac{1}{\mu - \lambda}\left(H_K + \frac{\lambda}{\mu}\left(\frac{2}{3}V_K - \frac{5}{6}H_K - \frac{1}{12}\right)\right), K \geq 2, 0 \leq \lambda \leq \mu. \tag{1.12}$$

If, respectively, the value of the characteristic parameter λ is described by the exponential distribution and the value of the characteristic parameter μ is described by the Erlang distribution of the second order: $A(x) = 1 - e^{-\lambda x}$, $B(x) = 1 - (1 + 2\mu x)e^{-2\mu x}$, $x \geq 0$, then expression (1.11) takes the form

$$E\left(W_{K,\max}\right) \approx \frac{1}{\mu - \lambda}\left(F_K + \frac{\lambda}{\mu}\left(-\frac{1}{12}H_K + \frac{2}{3}V_K - F_K + \frac{1}{6}\right)\right), \quad K \geq 2,\ 0 \leq \lambda \leq \mu. \tag{1.13}$$

Expressions (1.12) and (1.13) were obtained for a queuing network with h symmetric service stations of type $M|M|1$ by the H&LTIA method, with the following values of the key parameter: $\beta \in \{0; 0,5; 1\}$. Also in the article [52] the variants of adaptation of expressions (1.10) and (1.11) for cases when the values of the controlled parameters λ and μ are described by hyper exponential (exponential) and exponential (hyper exponential) distributions, respectively, are presented.

1.1.2 *Matrix-Geometric Method of Analysis of DC-Like Systems*

Assume that the analysis of the process of wireless information interaction of a set of end devices with a special purpose information system is carried out in an extended metric of characteristic parameters, which includes not only the already mentioned system response time but also the mathematical expectation and variance of synchronization time (the time delay between completion of service of the first and last subrequests of one incoming request). In this case, in contrast to the already considered methods, an adequate analysis tool is the theory of ordinal statistics [53, 54], the application of which also expands the range of potential distributions of such controlled parameters as the intensity of the flow of incoming requests λ and the value of the time of servicing of their corresponding subrequests μ. In the matrix-geometric method formulated in this paradigm, the studied process is considered as a quasi-process of «Birth and Death» (BD) [55–57]. In articles [6, 7], an algorithm for reducing the transition matrix of the studied Markov process to a block-diagonal matrix is proposed, which allows to compactly form a corresponding system of equilibrium equations concerning the required characteristic parameters and, now, solve it using known mathematical programming methods.

The application of the matrix-geometric method to estimate the response time of a DC-like system was investigated in articles [58–60]. In the article [58] models of functioning of a queuing network with K service stations of type $M|M|1$ with various intensities of service of subrequests μ_k and taking into account that

1. An incoming request is accepted by the system if at the time of its receipt the number of n_k active subrequests of the k-th type in the system is less than U_k: $n_k < U_k$, where U_k is an integer positive constant, $k = \overline{2, K}$. Otherwise, the incoming request is lost.
2. If the condition $n_i - n_j \neq U_{i,j}$ is not met for service stations i and j, $i \neq j, i, j \in K$, then the j-th service station is blocked until the i-th service station completes the active subrequest servicing process. Here n_k is the number of subrequests of the k-th type in the system, organized into a set $N = \{n_k\}$, $k = \overline{1, K}$ and $U_{i,j} \forall i, j$ is an integer positive constant.

The matrix-geometric method allows determining the stationary distribution of the number of active subrequests $\pi(N)$ in such a system. This distribution is the basis for the algorithmization of the process of calculating the response time distribution function $F_{W_{K,\max}}(x) = \sum_N \pi(N) F_{W^N_{K,\max}}(x), x > 0$, which depends on the conditional distributions $F_{W^N_{K,\max}}(x)$ under the condition that the system under study is in the state N. Condition 1 satisfies the conditional distributions $F_{W^N_{K,\max}}(x)$, defined as Erlang distributions, and condition 2—the distributions defined as a result of the analysis of the absorption process for subrequests. Also in the article [59], it is proved that conditions 1 and 2 determine the upper and lower limits of the response time of the studied system.

The article [60] investigates a queuing network with a Poisson flow of incoming requests and the same type of service stations, the value of the service time of subrequests which is described by the Cox distribution: $M|C_k|1$, $k = \overline{2, K}$. By analogy with the article [58], in the article [61] the studied system is approximated by two models:

1. A model in which all service stations except the first are characterized by a finite capacity.
2. Model with the introduced restriction on not exceeding the set threshold value by the differences between the lengths of queues of subrequests to any pair of service stations in the system under study.

In the article [62] the analysis of these models using a matrix-geometric method is carried out which results allowed to present matrices of probabilities of transition in a block-triangular form and to define stationary probabilities and, in turn, to define the distribution of time of response of the system and its moments. Finally, the article [63] proposes algorithms for determining the upper (model 1) and lower (model 2) limits of the response time of the queuing system with non-Poisson distribution of the intensity of the flow of incoming requests and non-exponential distribution of service time of the corresponding subrequests by multiple service stations. It is also known to use the matrix-geometric method to determine the mathematical expectation, variance and response time distribution function of ISM- and SM-systems [64], in particular, with a structure of service stations, the capacity of one of which is limited. In articles [62, 63] this method analysed the DC-like system with phase distribution of the intensity of the flow of incoming requests, and in the article [65]—the ISM-system with a developed policy of load sharing due to step-by-step planning of subrequests with service stations with separation of waiting and merging phases for related subrequests.

According to the source material [62–65], if $\{\xi_i\}$ is a finite sample defined on some probability space (Ω, F, P) such that $\omega \in \Omega : x_i = \xi_i(\omega)$, $i = \overline{1, K}$, then it is ordered in ascending order: $x_{(1)} \le x_{(2)} \le \ldots \le x_{(K)}$, and the re-indexed sequence $\{x_i\}, i = \overline{1, K}$, is called a variation series, and its elements are called ordinal statistics. Accordingly, the stochastic quantity $\xi_{(k)} : \xi_{(k)}(\omega) = x_{(k)}$ will be called the k-th ordinal statistics of the original sample, $k \in K$. By definition:

$$\xi_{(1)} = \min(\xi_1, \ldots, \xi_K), \xi_{(K)} = \max(\xi_1, \ldots, \xi_K). \tag{1.14}$$

Accordingly, in terms of the theory of ordinal statistics, the values of the response time $W_{K,\max}$ and $W_{K,\max}$ correspond to the stochastic values of $\xi_{(K)}$ and $\xi_{(1)}$, and the stochastic values of $\xi_{(k)}$, $k = \overline{2, K-1}$, characterize the time of subrequests in the system before they fall into the synchronization buffer.

The mathematical expectation of the response time can be determined by knowing the distribution of ordinal statistics $\xi_{(1)}$ and $\xi_{(K)}$, where the distribution function of the latter is a compatible distribution function of stochastic quantities $\xi_1, \ldots, \xi_K$:

$$F_{\xi(1)}(x) = P(\min(\xi_1, \ldots, \xi_K) \le x) = 1 - P(\xi_1 > x, \ldots, \xi_K > x), \tag{1.15}$$

$$F_{\xi(K)}(x) = P(\max(\xi_1, \ldots, \xi_K) < x) = P(\xi_1 < x, \ldots, \xi_K < x). \tag{1.16}$$

For a stochastic quantity, ξ_k we denote: $F_k(x)$ is the distribution function, $f_k(x)$ is the distribution density. If we accept the statement that stochastic quantities $\xi_1, \ldots, \xi_K$ are independent, then expressions (1.15), (1.16) can be rewritten:

$$F_{\xi(1)}(x) = 1 - P(\xi_1 > x, \ldots, \xi_K > x) = 1 - \prod_{i=1}^{K} (1 - F_i(x)), \tag{1.17}$$

$$F_{\xi(K)}(x) = P(\xi_1 < x, \ldots, \xi_K < x) = \sum_{i=1}^{K} F_i(x). \tag{1.18}$$

Accordingly, the n-th moment of the stochastic value of the response time can be determined as a result of integration:

$$E\left(W^n_{K,\min}\right) \approx E\left(\xi^n_{(1)}\right) = \int_0^\infty x^n f_{\xi_{(1)}}(x)dx, \; E\left(W^n_{K,\max}\right) \approx E\left(\xi^n_{(K)}\right) = \int_0^\infty x^n f_{\xi_{(K)}}(x)dx, \tag{1.19}$$

where respectively:

$$f_{\xi_{(1)}}(x) = \sum_{j=1}^{K} f_j(x) \prod_{i=1}^{K} \frac{1 - F_i(x)}{1 - F_j(x)}, f_{\xi_{(K)}}(x) = \sum_{j=1}^{K} \frac{f_j(x)}{F_j(x)} \prod_{i=1}^{K} F_i(x).$$

If we assume the homogeneity of the service stations of the studied DC-like system, it turns out that the independent stochastic quantities $\xi_1, \ldots, \xi_K$ are equally distributed, i.e. their distribution functions are the same: $F_k(x) = F(x), f_k(x) = f(x)$, $k = \overline{1, K}$. Considering this circumstance, we write

$$P(\xi_1 < x, \ldots, \xi_K < x) = (F(x))^K, P(\xi_1 > x, \ldots, \xi_K > x) = 1 - (1 - F(x))^K. \tag{1.20}$$

We use expressions (1.20) to redefine expressions (1.19):

$$E\left(\xi^n_{(1)}\right) = K \int_0^\infty x^n f(x)(1 - F(x))^{K-1} dx, \; E\left(\xi^n_{(K)}\right) = K \int_0^\infty x^n f(x)(F(x))^{K-1} dx. \tag{1.21}$$

Also, note that the mathematical expectation of the maximum of K is defined as [65]:

$$E\left(\xi^n_{(K)}\right) = \int_0^\infty x d(F(x))^K = \int_0^\infty \left(1 - (F(x))^k\right) dx. \tag{1.22}$$

Analysis of the response time of the studied DC-like system involves the calculation of the approximate value of integrals (1.21), (1.22) by appropriate computational methods, but for such distributions as exponential, hyper exponential, Erlang second order, Cox, known are analytical options for implementing the solution process [66, 67]. Note that the complexity of such calculations largely depends on the order of distribution of Erlang or Cox and the number of service stations in the system under study. When calculating the maximum in conditions of limited computing power, we can use the characteristic maximum [68]. Let us examine this computational process in more detail for each of these distributions.

Let the time of servicing of subrequests by service stations of the investigated DC-like system be generalized in the form of a set of exponentially distributed stochastic quantities ξ_k, $k = \overline{1, K}$, with distribution densities $f_k(x) = \lambda_k e^{-\lambda_k x}$, $x > 0$.

To determine the moments of the highest order of the maximum of K independent, exponentially distributed stochastic quantities, it is necessary to differentiate the Laplace–Stieltjes transform of the compatible distribution function of maximum $\pi_{\xi(K)}(s)$ by s the appropriate number of times with the appropriate equation of s to zero. In particular, as a result of performing a sequence of such operations at $K = 2$ we obtain the following distribution functions and accompanying expressions [65, 66]:

$$\begin{aligned} F_{\xi(2)}(x) &= F_1(x)F_2(x) = 1 - \exp(-\lambda_1 x) - \exp(-(\lambda_1 + \lambda_2)x), \\ f_{\xi(2)}(x) &= \frac{dF_2(x)}{d(x)} = f_1(x)F_2(x) + F_1(x)f_2(x) \\ &= \lambda_1 \times \exp(-\lambda_1 x) + \lambda_2 \exp(-\lambda_2 x) - (\lambda_1 + \lambda_2)\exp(-(\lambda_1 + \lambda_2)x), \\ \pi_{\xi(2)} &= \int_0^\infty f_2(x)e^{-sx}dx = \frac{\lambda_1}{s + \lambda_1} + \frac{\lambda_2}{s + \lambda_2} - \frac{\lambda_1 + \lambda_2}{s + \lambda_1 + \lambda_2}. \end{aligned} \tag{1.23}$$

Accordingly, the mathematical expectation of the maximum is defined as [66]:

$$E\left(\xi_{(2)}\right) = -\left.\frac{d\pi_2(s)}{ds}\right|_{s=0} = -\frac{1}{\lambda_1 + \lambda_2} + \frac{1}{\lambda_1} + \frac{1}{\lambda_2}.$$

In general (for K independent, exponentially distributed stochastic quantities), expression (1.20) is represented as [62]:

$$\pi_{\xi(K)}(s) = \sum_{i=1}^{K} (-1)^{i-1} \sum_{j=1}^{\binom{K}{i}} \sum_{m=1}^{j} \left(\sum_{l=1}^{i} \lambda_{(m+l-1)_K} \Big/ s + \sum_{l=1}^{i} \lambda_{(m+l-1)_K} \right), \tag{1.24}$$

where $\lambda_{(m+l-1)_K}$ is the result of the operation of sum mod K.

An analogue of expression (1.24) is a recurrent representation of the Laplace–Stieltjes transformation of the maximum of K independent, exponentially distributed stochastic quantities [69] with parameters $\lambda = (\lambda_1, \ldots, \lambda_K)$, $f_{\xi_{(K)}}(\lambda, x)$, $\pi_{\xi_{(K)}}(\lambda, s)$:

$$\left(s + \sum_{j=1}^{k} \lambda_j \right) \pi_{\xi_{(k)}}(\lambda, s) = \sum_{j=1}^{k} \pi_{\xi_{(k-1)}} \lambda_j \left(\lambda_{/j}, s\right), \ 1 \le k \le K,$$

where $\pi_0(0, s) = 1$ $\lambda_{/j}$ is the set of λ from which the element λ_j is excluded: $\lambda_{/j} = (\lambda_1, \ldots, \lambda_{j-1}, \lambda_{j+1}, \ldots, \lambda_K)$. Accordingly, the moment of the n-th order of a maximum of K independent, exponentially distributed stochastic quantities is defined as

$$M_K(\lambda, n) = \frac{nM_K(\lambda, n-1) + \sum_{j=1}^{K} M_{K-1}(\lambda_{/j}, n)}{\sum_{j=1}^{K} \lambda_j}, \tag{1.25}$$

where $K \geq 1$, $M_0(0, n) = 0 \forall n \geq 1$, $M_K(\lambda, 0) = 1 \forall K \geq 0$.

Let us investigate another important characteristic parameter of DC-like systems—the synchronization time (meaningful analogue of the theory of ordinal statistics—the scope [70, 71]). This parameter is equal to the time interval from the moment when the first to the last subrequests of one incoming request enters the synchronization buffer. In other words, the synchronization time is equal to the difference between the maximum and minimum values from the set of values of the residence time of the subrequests of the corresponding input request in a DC-like system:

$$E(W_K) = E(W_{K,\max}) - E(W_{K,\min}). \tag{1.26}$$

In [72], the task of minimizing the characteristic parameter (1.26) for a DC-like system with two homogeneous service stations was first formulated and numerically solved. In [73], estimates of the mathematical expectation of the synchronization time of such systems in the composition of K inhomogeneous service stations and the variance of this characteristic parameter for a DC-like system with K homogeneous service stations ($\lambda_k = \lambda \forall k = \overline{1, K}$) were presented:

$$D(W_K) \approx \frac{1}{\lambda^2} \sum_{i=1}^{K-1} \frac{1}{i^2}.$$

In the theory of ordinal statistics [70], there is an expression for determining the scale distribution function of a normally distributed stochastic quantity:

$$F_{W_K}(x) \approx F_{\xi_{(K)} - \xi_{(1)}}(x) = \sum_{i=1}^{K} \int_0^{\infty} f_i(y) \prod_{j=1, j \neq i}^{K} (F_j(y + x) - F_j(y))dy, \tag{1.27}$$

and an expression for determining the scale variance of a set of stochastic quantities with power K with distribution function $F(x)$:

$$D(W_K) \approx D(\xi_{(K)} - \xi_{(1)}) =$$
$$= 2 \int_{-\infty}^{\infty} (1 - F^K(y) - (1 - F(x))^K + (F(y) - F(x))^K)dxdy - (E(W_K))^2. \tag{1.28}$$

Expression (1.28) can be used to estimate the variance of the synchronization time of a DC-like system with K inhomogeneous service stations with the non-exponential distribution of stochastic quantities generalized by the set λ.

As noted in [35], if the values of the time of subrequests servicing by service stations of DC-like system are described by the Erlang distribution (ε_{l_k}) of the l_k-th order with the following distribution function $F(x)$ and distribution density $f(x)$ such that

$$F(x) = 1 - e^{-\lambda_k x} \sum_{i=0}^{l_k - 1} \frac{(\lambda_k x)^i}{i!}, f(x) = \frac{\lambda_k (\lambda_k x)^{l_k - 1}}{(l_k - 1)!} e^{-\lambda_k x},$$

where $\lambda_k, x \geq 0$, $l_k > 0$, then the mathematical expectation of stochastic quantities l_k / λ_k at $K = 2$ is defined as

$$E(W_{2,\max}) \approx \sum_{k=1}^{2} \frac{l_k}{\lambda_k} - \sum_{m=0}^{l_1 - 1} \sum_{n=0}^{l_2 - 1} \binom{m+n}{m} \frac{\lambda_1^m \lambda_2^n}{(\lambda_1 + \lambda_2)^{m+n+1}},$$

and at $K \geq 2$:

$$E(W_{K,\max}) \approx \int_0^\infty \left(1 - e^{-\lambda_k x} \sum_{i=0}^{l_k - 1} \frac{(\lambda_k x)^i}{i!}\right) dx.$$

1.1.3 Method of Distribution of Extreme Values as a Tool for Analysis of DC-Like Systems

A functional analogue of the matrix-geometric method for determining the maximum of independent, normally distributed stochastic characteristic parameters of a DC-like system is the method of approximation by the distribution of extreme values [74]. This distribution is the limiting distribution of the extreme (maximum $\xi_{(K)}$ or minimum $\xi_{(1)}$) value of normally distributed continuous stochastic quantities at $K \to \infty$. Of course, the distribution $\xi_{(K)}$ obtained by this method at $K \to \infty$ will be degenerated. To obtain a nondegenerate distribution, investigate the linear transformation $\xi_{(K)}$ with coefficients depending on the sample size from the continuous distribution.

There are three classes of extreme distributions [75]:

$$F_{\xi_{(K)}}(x) = e^{-e^{-\frac{x-a}{\theta}}}, \tag{1.29}$$

$$F_{\xi_{(K)}}(x) = \begin{cases} 0 \forall x < \alpha, \\ e^{-\left(\frac{x-\alpha}{\theta}\right)^{-n}} \forall x \geq \alpha, \end{cases} \tag{1.30}$$

$$F_{\xi_{(K)}}(x) = \begin{cases} e^{-\left(\frac{x-\alpha}{\theta}\right)^{n}} \forall x < \alpha, \\ 0 \forall x \geq \alpha, \end{cases} \tag{1.31}$$

where $\alpha, \theta, n > 0$ are the controlled parameters.

Distributions (1.29), (1.30), and (1.31) are classified as Gumbel distribution, Frechet distribution and Weibull distribution, respectively. The last two classes of distributions are reduced to the Gumbel distribution using the transformations $\left(\ln\left(\xi_{(K)} - \alpha\right)\right)$ and $\left(-\ln\left(\xi_{(K)} - \alpha\right)\right)$, respectively. So, let's focus on the Gumbel distribution. Density $f_{\xi_{(K)}}(x)$, mathematical expectation $E\left(\xi_{(K)}\right)$, and variance $D\left(\xi_{(K)}\right)$ of this distribution are defined as [75]:

$$\begin{aligned} f_{\xi_{(K)}}(x) &= \frac{1}{\theta} e^{-\frac{x-\alpha}{\theta}} e^{-e^{-\frac{x-\alpha}{\theta}}}, \\ E\left(\xi_{(K)}\right) &= \alpha + \gamma\theta, \\ D\left(\xi_{(K)}\right) &= \frac{1}{6}\pi^2\theta^2, \end{aligned} \tag{1.32}$$

where γ is the Euler–Mascheroni constant.

Variants of representation of expressions (1.29), (1.31), obtained at $\alpha = 0$ and $\theta = 1$:

$$F_{\xi_{(K)}}(x) = e^{-e^{-x}}, \quad f_{\xi_{(K)}}(x) = e^{-x-e^{-x}}, \tag{1.33}$$

are known as the standard form of Gumbel distribution.

In [76], the necessary and sufficient conditions for the convergence of the studied general distribution $F(x)$ to one of the classes of distributions (1.29), (1.30), or (1.31) at large (small) values relative to the maximum (minimum) values of stochastic quantities are formulated. Moreover, with a single general distribution, the limit distributions for extreme values can be of different types.

The distributions that coincide with the Gumbel distribution include normal, exponential, and logical distributions. Distributions that coincide with the Frechet distribution include the Cauchy distribution. Distributions that coincide with the Weibull distribution include distributions centred on the upper part of the numerical axis [77]. In particular, in [78] it was shown that the distribution of the maximum value of $\xi_{(K)}$ with increasing sample size of stochastic quantities, which are described by an exponential distribution with the characteristic parameter λ, approaches the standard form of the Gumbel distribution:

$$\xi_{(K)} \approx \ln K + \varsigma, E\left(\xi_{(K)}\right) \approx \ln K + \gamma, D\left(\xi_{(K)}\right) \approx D(\varsigma) \approx \frac{\pi^2}{6},$$

where ς is a stochastic quantity characterized by the Gumbel distribution with distribution function [79], mathematical expectation $E(\varsigma) = \gamma$, and variance $D(\varsigma) = \pi^2/6$.

In [79], for the case when each ordinal statistic is characterized by a standard Humbell distribution (1.33) for extreme values, it is shown that the mathematical expectation of the maximum is determined according to the expression $E(\xi_{(K)}) = \alpha + \theta\gamma + \theta \ln K$.

We summarize the mathematical apparatus presented in [80] for the approximation of the moment of the maximum of stochastic quantities, which are described by an arbitrary distribution. We investigate the stochastic quantity $\xi = \max\{\xi_k\}$, $k = 1, 2$, where ξ_k is a positive independent stochastic quantity with the distribution function $F_k(x)$ and the Laplace–Stieltjes transformation $\pi_k(s)$ of this function.

We define the mathematical expectation ξ as

$$E(\xi) = E(\xi_1) + P(\xi_2 > \xi_1) \cdot E(\xi_2 - \xi_1 | \xi_2 > \xi_1) =$$
$$= E(\xi_1) + \int_0^\infty F_1(x) dF_2(x) \cdot E(\xi_2 - \xi_1 | \xi_2 > \xi_1).$$

Let μ_1, μ_2 be the first and σ_1, σ_2 be the second moments of the stochastic quantities ξ_1, ξ_2, respectively. Then, if ξ_1 is distributed exponentially, then

$$E(\xi_2 - \xi_1 | \xi_2 > \xi_1) = \frac{\sigma_2}{2\mu_2},$$

The final version of the formalization of the desired approximation is as follows:

$$E(T) = \mu_1 + \frac{\sigma_2 \pi_1\left(\mu_2^{-1}\right)}{\mu_2}. \tag{1.34}$$

If both stochastic quantities ξ_1 and ξ_2 are distributed exponentially, then the approximation of the moment of the maximum of these stochastic quantities in the form of expression (1.34) is an exact solution.

In the general case (when $K > 2$), it is necessary to approximate the Laplace–Stieltjes transformation of the maximum distribution function of $(K - 1)$ stochastic quantities. Assume that expression (1.25) can be used to determine the moments of non-exponential stochastic quantities. Denote by $M(K, \lambda, \sigma)$ the mathematical expectation of the maximum of K stochastic quantities with the first moment $\mu = (\mu_1, \ldots, \mu_K)$, $\lambda = \left(\lambda_1 = \frac{1}{\mu_1}, \ldots, \frac{1}{\mu_K} = \lambda_K\right)$ and the second moment $\sigma = (\sigma_1, \ldots, \sigma_K)$. Then the value $M(K, \lambda, \sigma)$ can be expressed recurrently:

$$M(K, \lambda, \sigma) = \frac{1}{2}\lambda_i \sigma_i \pi_{\xi_{(k-1)}}\left(\lambda_{/i}, \lambda_i\right) + \frac{1}{k}\sum_{i=1}^{k} M\left(k - 1, \lambda_{/i}, \sigma_{/i}\right), k = \overline{2, K}, \tag{1.35}$$

where $M(1, \lambda_1, \sigma_1) = \lambda_1^{-1}$.

If all the studied stochastic quantities are described by the same distribution law with $\lambda_i = \lambda$, $\sigma_i = \sigma$, $\forall 1 \leq i \leq K$, then

$$M(K, \lambda, \sigma) = \frac{1}{\lambda} + \frac{1}{2}\sigma\lambda(H_k - 1).$$

In [77, 78], the results of simulation modelling with the help of the above mathematical apparatus are shown. Distributions such as the Erlang distribution of 2–4 orders ($\varepsilon_2, \varepsilon_2, \varepsilon_5$) and the Pareto distribution with the distribution function $F_p(x) = 1 - \alpha(x+\gamma)^{-\beta}$, $\beta = \{4, 5\}$ were studied. Analysis of the distribution function $F_p(x)$ showed that for the first two moments to be finite, it is necessary that the condition $\beta > 2$ be satisfied: $\alpha = \gamma^{\beta}$, $\gamma = \beta - 1$, so that $m = 1$, $M = 2 + 2/(\beta - 2)$. The approximation error of expressions (1.21) and (1.35) for the distribution ε_2 remains small enough despite the increase in the value of K, but for the distributions $\varepsilon_3, \varepsilon_4$ at $K \geq 16$ the approximation error begins to increase rapidly. At the same time, both equations provide relatively accurate results for the Pareto distribution, but the first expression predominates.

Denote the mathematical expectation and variance of the general distribution as μ and σ^2, respectively. In [81] based on these characteristic parameters the condition $E(\xi_{(K)}) \leq \mu - \sigma\frac{K-1}{\sqrt{2K-1}} \leq E(\xi_{(1)})$ is formulated, taking into account which, to determine the maximum of K independent, equally distributed stochastic quantities $\xi_{(K)}$, an approximating expression

$$E(\xi_{(K)}) \approx \mu + \sigma G(K) \tag{1.36}$$

is proposed, where

$$G(K) \leq \frac{K-1}{\sqrt{2K-1}}. \tag{1.37}$$

For example, for the case of the exponential distribution $G(K) = H_K - 1$, and for uniform distribution—$G(K) = \sqrt{3}(K-1)/(K+1)$. If the mathematical expectation is zero, and the standard deviation is one, then $E(\xi_{(K)}) \approx G(K)$. Expression (1.37) becomes equality for independent, uniformly distributed stochastic quantities (ξ_k), $k = \overline{1, K}$, with the distribution function and distribution density of the form

$$F(x) = \left(\frac{1+bx}{K}\right)^{\frac{1}{K-1}}, \; f(x) = \frac{b}{K}\left(\frac{1+bx}{K}\right)^{\frac{1}{K-1}},$$

where $-\frac{\sqrt{2K-1}}{K-1} \leq x \leq \sqrt{2K-1}$, $b = \frac{K-1}{\sqrt{2K-1}}$.

For a symmetric distribution $F(x) = 1 - F(-x)$ we can write

$$E(\xi_{K,\max}) \leq \frac{K}{2}\sqrt{\frac{2-\left(1-1/\frac{2K-2}{K-1}\right)}{2K-1}}.$$

To approximate the expected maximum of independent, uniformly distributed stochastic quantities with the normal distribution law in expression (1.36) use $G(K) = \sqrt{2\ln(K)}$:

$$E(\xi_{(K)}) \approx \mu + \sigma\sqrt{2\ln(K)},$$

but a more accurate expression was presented in [79]:

$$E(\xi_{(K)}) \approx \mu + \sigma\left(\sqrt{2\ln(K)} - \frac{\ln(\ln(K)) - \ln(4\pi) - 2\gamma}{2\sqrt{2\ln(K)}}\right). \quad (1.38)$$

If we take into account the shift $0,1727K^{-0,2750}$ in expression (1.38), we obtain an expression for estimating the variance of the approximation of the expected maximum of independent, uniformly distributed stochastic quantities with the normal distribution law:

$$D(\xi_{(K)}) \approx \frac{1,64492\sigma^2}{2\ln(K)}.$$

In [82], based on expression (1.36), an expression is formulated for the approximate determination of the stochastic value of the response time:

$$E(W_{K,\max}) \approx E(W_{1,\max}) + F_K\sigma_{W_{1,\max}}(\rho)\alpha_K(\rho),$$

where $E(W_{1,\max})$ and $\sigma_{W_{1,\max}}(\rho)$ are, respectively, the mathematical expectation and the standard deviation of the response time of the queuing system type $M|G|1$ with a loading coefficient ρ.

If in such a system the values of subrequests servicing time for all K service stations are described by an arbitrary distribution law $B(x)$, then the value of $E(W_{1,\max})$ can be defined as [82]:

$$E(W_{1,\max}) = b + \frac{\lambda b^{(2)}}{2(1-\rho)},$$

where $b = \int\limits_0^\infty x dB(x)$.

The value of the parameter F_K from the above expression for determining $E(W_{K,\max})$ is defined as

$$F_K = \frac{b_{K,\max} - b}{\sqrt{b^{(2)} - b^2}}$$

at $\rho = 0$, $\alpha_K(\rho) = 1$, where $b_{K,\max} = \int\limits_0^\infty \left(1 - B^K(x)\right)dx$ is the mathematical expectation of the maximum of K independent, uniformly distributed stochastic quantities with the distribution $B(x)$. In particular, for the exponential distribution $F_K = H_K - 1$ and for the distribution of extreme values $F_K = \sqrt{6}\ln(K)\big/\pi$ [83].

In [84], as a result of generalization of the above analytical expressions and the results of simulation at a considered distribution of response time, the expression for determining $\alpha_K(\rho)$ was proposed:

$$\alpha_K(\rho) = \frac{E\left(W_{K,\max}\right) - E\left(W_{1,\max}\right)}{F_K \sigma_{W_{1,\max}}}.$$

For example, for the DC-like system with two service stations of the type $M|M|1$ mentioned in expression (1.28), it turns out that $\alpha_2(\rho) = 1 - \rho\big/4$.

Therefore, we consider concepts based on the matrix-geometric method and the method of ordinal statistics for approximate analysis of mathematical expectation and variance of response time and synchronization time for DC-like system with $K > 2$ service stations, the time stay of subrequests in which is described by the corresponding distribution law.

1.2 Model of Functioning of a Specimen of DC-Like Systems in the Conditions of Probable Cyber Threats

Let us investigate the model of a DC-like system in the form of a queuing system with independent streams of incoming requests of the first and second types. Incoming requests of the first type have absolute priority over incoming requests of the second type. At the completion of service for a type-1 request, a procedure for clearing the system's service station of type-2 requests may be initiated with a certain probability. In specialized information sources, this procedure is known as an update [85, 86]. The development of this concept is the concept of generalized update, which corresponds to the procedure of incomplete clearing of the system's service station from requests of the second type at the time of completion of the service of the request of the first type. When implementing a generalized update, the number of lost requests of the second type as a result of the clearing is characterized by a stochastic value with a certain distribution. The obvious functional application of the generalized update is the active control of the queuing system. The control strategy is implemented through the settings and selection of the moment of implementation of the generalized update procedure. The simplest is the threshold strategy [87], but more efficient RED-like stochastic algorithms are known [88].

Based on the just mentioned developments on the description of queuing systems with the update, we propose a model of the process of functioning of the base station of a centralized, wireless network in the conditions of probable cyber threats. The studied information system must effectively control the flow of incoming requests of the second type of any intensity while remaining ergodic. The queuing system is a model of a typical information system that serves ordinary incoming requests from the user's end devices (requests of the second type). The request of a first type corresponds to a cyber threat. The information security subsystem can neutralize it (the request of the first type is lost before its service) or pass (the request of the first type after its service partially (generalized update) or completely (update) clears the service station with a queue of requests of the second type). The research of such a system involves the analytical formalization of the condition of its ergodicity, the distribution of the stochastic number of requests in the system, the probability of its downtime, and so on.

Therefore, suppose that the input of a single-line queuing system with a service station of unlimited capacity is directed to two independent Poisson streams of input requests of the first and second types with intensities λ_1 and λ_2 respectively. The values of the service time of requests of both types are determined by stochastic, exponentially distributed values of μ_1 and μ_2, respectively. Incoming requests of the first type are an absolute priority, i.e. the request of the second type is selected for service only if there are no requests of the first type in the service station. Moreover, if the system's service station receives a request of the first type during the service of the request of the second type, this process is terminated, the request of the second type is fed to the beginning of the service station and the service of the request of the first type begins. The requests of the same type are served in the order of receipt. At the end of the service of the request of the first type with a probability of $p(q)$ may occur (not occur) the clearing of a service station from all received requests of the second type, $p + q = 1$. Hereinafter, we will call such a mechanism a stochastic update.

We identify the process of functioning of such a system as Markov, $X(t) = \{v_1(t), v_2(t)\}$, where $v_1(t)$ and $v_2(t)$ are functions with a discrete set of states (see the UML state diagram presented in Fig. 1.4), which characterize the number of requests of the first (i) and second (j) types in the service station.

Since the flows of incoming requests of both types are independent, the function $v_1(t)$ will be characterized by a queuing system of type $M|M|1$, the condition of ergodicity of which is: $\rho_1 < 1$, $\rho_i = \lambda_i / \mu_i$, $i = 1, 2$. In this context, the value of the function $v_2(t)$ does not go to infinity, because, according to the functioning scenario of the system under study, at stochastically defined (with finite mathematical expectation) time intervals, the service station is cleared of requests of the second type. As a result, the condition of ergodicity of the studied system is a condition $\rho_1 < 1$, which will be considered to be fulfilled by default in the future.

Let us denote the stationary probabilities that there are i requests of the first and j requests of the second types in the system's service station by the identifier $p_{i,j}$, $i, j \geq 0$. The stationary distribution of the probability of realization of such an event exists and is characterized by the following system of equilibrium equations:

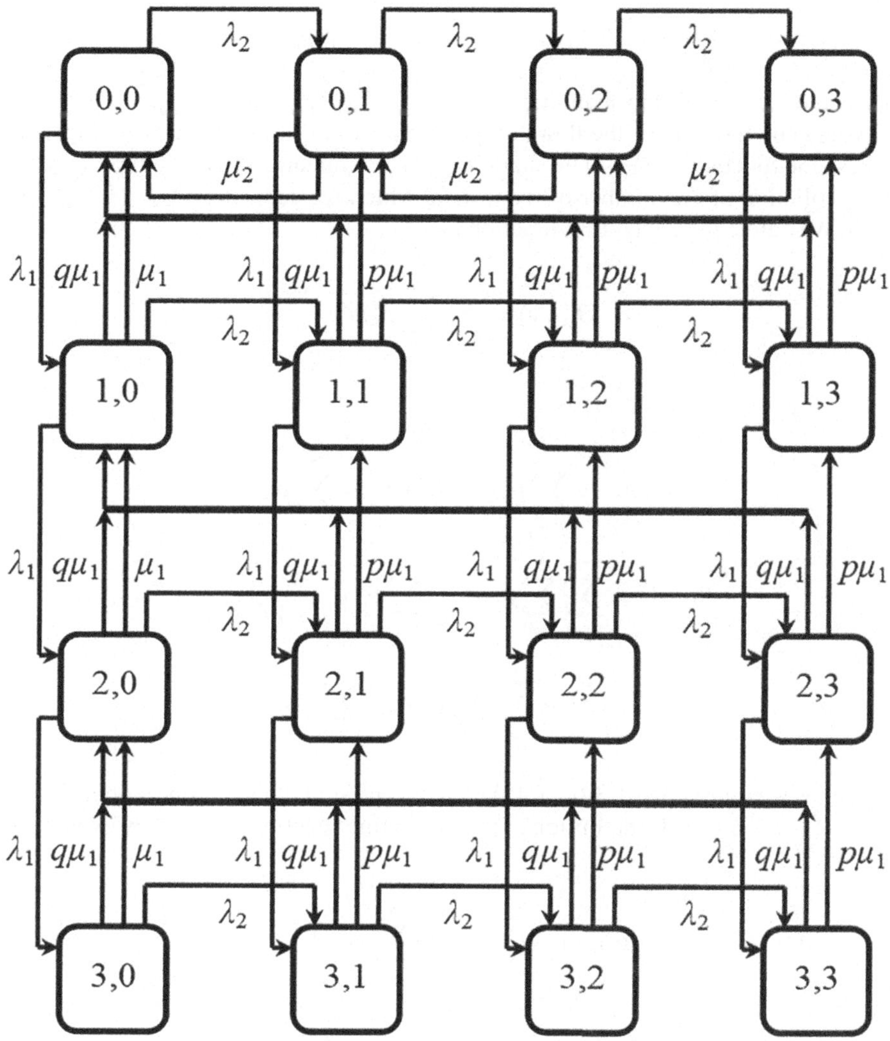

Fig. 1.4 UML state diagram of the process $X(t) = \{v_1(t), v_2(t)\}$

$$(\lambda_1 + \lambda_2)p_{0,0} = \mu_1 p p_{1,0} + \mu_2 p_{0,1} + \mu_1 q \sum_{j=0}^{\infty} p_{1,j}; \tag{1.39}$$

$$(\lambda_1 + \lambda_2 + \mu_1)p_{i,0} = \lambda_1 p_{i-1,0} + \mu_1 p p_{i+1,0} + \mu_1 q \sum_{j=0}^{\infty} p_{i+1,j}, \ i \geq 1; \tag{1.40}$$

$$(\mu_2 + \lambda_1 + \lambda_2)p_{0,j} = \lambda_2 p_{0,j-1} + \mu_2 p_{0,j+1} + \mu_1 p p_{1,j}, \ j \geq 1; \tag{1.41}$$

$$(\mu_1 + \lambda_1 + \lambda_2)p_{i,j} = \lambda_1 p_{i-1,j} + \lambda_2 p_{i,j-1} + \mu_1 p p_{i+1,j}, \; i, j \geq 1. \tag{1.42}$$

Denote the marginal probabilities as $p_{i,\cdot} = \sum_{j=0}^{\infty} p_{i,j}$, $p_{\cdot j} = \sum_{i=0}^{\infty} p_{i,j}$, $i, j \geq 0$.

As already mentioned, the flows of incoming requests of both types are independent, so the function $v_1(t)$ will be characterized by a queuing system of type $M|M|1$, the distribution of the number of requests in which we define as $p_{i,\cdot} = (1-\rho_1)\rho_1^i$, $i \geq 0$. We define the generating function as

$$B(u, v) = \sum_{i=1}^{\infty}\sum_{j=0}^{\infty} p_{i,j} u^i v^j. \tag{1.43}$$

Interpret expression (1.43) as

$$B(u, 0) = \sum_{i=0}^{\infty} p_{i,0} u^i, \; B(0, v) = \sum_{j=0}^{\infty} p_{0,j} u^j,$$

$$B(u, 1) = \sum_{i=0}^{\infty}\sum_{j=0}^{\infty} p_{i,j} u^i = \sum_{i=0}^{\infty} p_{i,\cdot} u^i = \frac{1-\rho_1}{1-\rho_1 u},$$

$$B(1, v) = \sum_{i=0}^{\infty}\sum_{j=0}^{\infty} p_{i,j} v^j = \sum_{j=0}^{\infty} p_{\cdot j} v^j.$$

Multiply expressions (1.39)–(1.42) by $u^i v^j$. The sum of the products obtained for all values of i and j is the definition of the generating function declared by expression (1.43):

$$\begin{aligned}
&(\lambda_1 + \lambda_2 + \mu_1)B(u, v) + (\mu_2 - \mu_1)\sum_{j=1}^{\infty} p_{0,j} v^j - \mu_1 p_{0,0} \\
&= \mu_1 \frac{q}{u} \sum_{j=0}^{\infty} p_{1,j} u + \mu_1 p p_{1,0} + \mu_2 p_{0,1} + \mu_1 q \sum_{i=1}^{\infty}\sum_{j=0}^{\infty} p_{i+1,j} u^i + \\
&+ \mu_1 p \sum_{j=1}^{\infty} p_{1,j} u^j + \lambda_2 \sum_{j=1}^{\infty} p_{0,j-1} v^j + \lambda_1 \sum_{i=1}^{\infty} p_{i-1,0} u^i + \\
&+ \mu_1 p \sum_{i=1}^{\infty} p_{i+1,0} u^i + \sum_{i=1}^{\infty}\sum_{j=1}^{\infty} p_{i+1,j} u^i v^j + \mu_1 q \sum_{i=1}^{\infty}\sum_{j=1}^{\infty} p_{i-1,j} u^i v^j + \\
&+ \lambda_2 \sum_{i=1}^{\infty}\sum_{j=1}^{\infty} p_{i,j-1} u^i v^j + \mu_2 \sum_{j=1}^{\infty} p_{0,j+1} v^j + \mu_1 p.
\end{aligned} \tag{1.44}$$

Simplifying expression (1.44) we obtain a compact formula for calculating the generating function (1.43):

$$B(u,v) = \frac{(\mu_1(u-p) + \mu_2 u(v-1))B(0,v)}{\lambda_1 uv(1-u) + \lambda_2 uv(1-u) + \mu_1 v(u-p)} + \\ + \frac{\mu_1 qv\left(B(u,1) - p_{0,\cdot}\right) + \mu_2 p_{0,0} u(v-1)}{\lambda_1 uv(1-u) + \lambda_2 uv(1-u) + \mu_1 v(u-p)}. \tag{1.45}$$

Based on expression (1.45) we formulate the distribution for the stochastic value of the number of requests of the second type in the system's service station:

$$p_{\cdot j} = \frac{1}{j!}\frac{\partial^j B(1,v)}{\partial v^j}\bigg|_{v=0}. \tag{1.46}$$

In turn, taking into account expression (1.46), we formulate a distribution for the stochastic value of the number of requests of the first and second types in the system's service station:

$$p_{i,j} = \frac{1}{i!j!}\frac{\partial^{i+j} B(u,v)}{\partial u^i \partial v^j}\bigg|_{u=v=0}. \tag{1.47}$$

Express the roots of the quadratic trinomial (1.45) concerning the variable u:

$$u_{1,2} = u_{1,2}(v) = \frac{1}{2\lambda_1}(\lambda_1 + \lambda_2(1-v) + \mu_1 \pm \\ \pm\sqrt{(\lambda_1 + \lambda_2(1-v) + \mu_1)^2 - 4\lambda_1\mu_1 p}\Big). \tag{1.48}$$

Prove: if $\rho_1 = \lambda_1 / \mu_1 < 1$, then $0 < u_2(v) < 1$. Consider the function

$$\lambda_1 u(1-u) + \lambda_2 u(1-v) + \mu_1(u-p) \tag{1.49}$$

from the denominator of expression (1.45). For $u = 0$ we have: $-\mu_1 p < 0$, and for $u = 1$: $\lambda_2(1-v) + \mu_1(1-p) > 0, 0 < p < 1$. In general, the graph of this function is a parabola visualized in Fig. 1.5.

From Fig. 1.5 graph, it becomes obvious that the root of function (1.48) in the interval (0, 1) exists. In addition, this root is the smallest of the roots of this square trinomial.

The generating function $B(u, v)$ is continuous for $u, v \in [0, 1]$. Given this, when substituting the values of (u_2, v), $0 < u_2 < 1$, in expression (1.45), both the numerator and denominator of the latter become equal to zero:

$$(\mu_1 v(u_2 - p) - \mu_2 u_2(v-1))B(0,v) + \\ + \mu_1 qv\left(B(u_2,1) - p_{0,\cdot}\right) + \mu_2 p_{0,0} u_2(v-1) = 0, \tag{1.50}$$

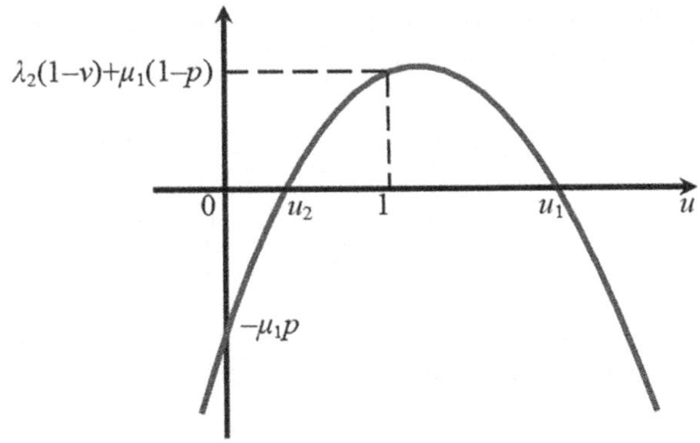

Fig. 1.5 Graph of function (1.49)

where $B(u_2, 1) = (1 - \rho_1) / (1 - \rho_1 u_2)$, $p_{0,\cdot} = 1 - \rho_1$.

Taking into account expression (1.50), we can formulate an expression for calculating $B(0, v)$:

$$B(0, v) = \frac{(1 - \rho_1 u_2)\mu_2 u_2 (1 - v) p_{0,0} + \mu_1 q \rho_1 u_2 (\rho_1 - 1) v}{(\mu_1 v (u_2 - p) - \mu_2 u_2 (v - 1))(1 - \rho_1 u_2)}, \tag{1.51}$$

all variables in which are already defined except for the probability of downtime $p_{0,0}$.

Let us focus on the denominator of expression (1.51). Since $v \in [0, 1]$, at the boundaries of this interval, for $v = 0$, the expression acquires the value of $\mu_2 u_2 > 0$, but the sign of the value of $\mu_1 v (u_2 - p)$, to which expression (1.51) coincides at the boundary point $v = 1$, must be determined. From the denominator of expression (1.45) we obtain

$$\mu_1 (u_2 - p) + \lambda_1 u_2 (1 - u_2) + \lambda_2 u_2 (1 - v) = 0,$$

where $\lambda_1 u_2 (1 - u_2) > 0$ i $\lambda_2 u_2 (1 - v) > 0$.

Accordingly, we consider $\mu_1 (u_2 - p) < 0$. That is, there is a value $v^* \in (0, 1)$ at which the denominator of expression (1.51) becomes equal to zero, respectively, and the numerator (1.51) under the following conditions must be equal to zero:

$$\mu_1 q \rho_1 u_2 (v^*) (\rho_1 - 1) v^* + (1 - \rho_1 u_2 (v^*)) \mu_2 u_2 (v^*) (1 - v^*) p_{0,0} = 0. \tag{1.52}$$

Assuming that $0 < u_2(v) < 1$, from expression (1.52) we obtain

$$p_{0,0} = \frac{\lambda_1 q (1 - \rho_1) v^*}{\mu_2 (1 - \rho_1 u_2 (v^*))(1 - v^*)}. \tag{1.53}$$

Finally, having determined by expression (1.53) the probability of downtime $p_{0,0}$, we can calculate all the stationary characteristics and probabilities of the system under study.

We show that for $p \to 0$: $p_{0,0} \to p_{0,0}^{(0)} = \frac{\lambda_1(1-\rho_1)}{\mu_2}\frac{z_2}{1-z_2}$, $v^* \to z_2$, $u_2 \to 0$, where

$$z_2 = \frac{1}{2\lambda_2}\left(\mu_1 - \sqrt{(\mu_1 + \lambda_1 + \lambda_2)^2 - 4\lambda_2\mu_2} + \lambda_1 + \lambda_2\right), 0 < z_2 < 1. \quad (1.54)$$

As already shown, at $p \to 0$ for all values v is valid $u_2 \to 0$. We obtain the asymptotic distribution of the expression for u_2 as a result of its expansion into the Taylor series at $p \to 0$:

$$u_2 = \frac{\mu_1 p}{\lambda_1 + \lambda_2(1+v) + \mu_1} + o(p), \quad (1.55)$$

where

$$p - u_2 = p\frac{\lambda_1 + \lambda_2(1-v)}{\lambda_1 + \lambda_2(1-v) + \mu_1} + o(p). \quad (1.56)$$

We equate the denominator of expression (1.51) to zero and, finally, we obtain

$$\frac{v}{1-v} = \frac{\mu_2 u_2}{\mu_1(p - u_2)}. \quad (1.57)$$

In the right part of the expression (1.57), we substitute asymptotic expressions (1.55) and (1.56). After simplification we get

$$\frac{v}{1-v} = \frac{\mu_2}{\lambda_1 + \lambda_2(1-v)} + o(1). \quad (1.58)$$

For $p \to 0$, based on expression (1.58) we obtain the expression

$$\lambda_2 v^2 - (\lambda_1 + \lambda_2 + \mu_2)v + \mu_2 = 0,$$

the solution of which in the analytical form exactly corresponds to expression (1.54). Accordingly, for $v^* \to z_2$ and $u_2 \to 0$ we have $p_{0,0} \to p_{0,0}^{(0)}$.

Now let us investigate the opposite boundary case: $p \to 1$. We show that under such boundary conditions $v^* \to 1$, $u_2 \to 1$, and for $\rho_1 + \rho_2 < 1$: $p_{0,0} \to p_{0,0}^{(1)} = 1 - \rho_1 - \rho_2$. Denote by $\varepsilon = 1 - v$, then $\varepsilon \to 0$, $q \to 0$. Using the asymptotic expansion u_2 using the Taylor formula, we obtain

$$u_2 = 1 - \varepsilon\frac{\lambda_2}{\mu_1 - \lambda_1} - q\frac{\mu_1}{\mu_1 - \lambda_1} + o(\varepsilon) + o(q). \quad (1.59)$$

From expression (1.59) we obtain an analogue of expression (1.56):

$$p - u_2 = \varepsilon\frac{\lambda_2}{\mu_1 - \lambda_1} + q\frac{\lambda_1}{\mu_1 - \lambda_1} + o(\varepsilon) + o(q).$$

From expression (1.57) for $p \to 1$, $v \to 1_2$, $u_2 \to 1$ follows $1 - v \sim \mu_1(p - u_2)/\mu_2$.

Accordingly, at $\varepsilon, q \to 0$:

$$\varepsilon \sim \frac{\mu_1}{\mu_2}\left(-q\frac{\mu_1}{\mu_1 - \lambda_1} + \varepsilon\frac{\lambda_2}{\mu_1 - \lambda_1}\right) = \frac{q\lambda_1\mu_1}{\mu_1\mu_2 - \lambda_2\mu_1 - \lambda_1\mu_2}. \quad (1.60)$$

Based on expression (1.60) we obtain

$$p_{0,0} \sim \frac{\lambda_1 q}{\mu_2 \varepsilon} \to \frac{\mu_1\mu_2 - \lambda_1\mu_2 - \lambda_2\mu_1}{\mu_1\mu_2} = 1 - \rho_1 - \rho_2 = p_{0,0}^{(1)}.$$

which was necessary to find.

Let us determine the mathematical expectation of the number of requests of the second type in the system under study as

$$N_2 = \sum_{j=0}^{\infty} j p_{\cdot j} = \left.\frac{\partial B(1, v)}{\partial v}\right|_{v=1}. \quad (1.61)$$

Let's perform successive transformations of the right part of the expression (1.61):

$$\left.\frac{\partial B(1, v)}{\partial v}\right|_{v=1} =$$
$$= \frac{\lambda_1 u_2(1) q(1 - \rho_1) + \mu_2 u_2(1)(1 - \rho_1 u_2(1)) p_{0,0} + u_2'(1)\lambda_1 q(1 - \rho_1)}{\mu_1(p - u_2(1))(1 - \rho_1 u_2(1))} -$$
$$- \frac{\left(\mu_1(p - u_2(1)) + \mu_2 u_2(1) - u_2'(1)\mu_1\right)\lambda_1 u_2(1) q(1 - \rho_1)}{\mu_1^2(p - u_2(1))(1 - \rho_1 u_2(1))} +$$
$$+ \frac{u_2(1) q(1 - \rho_1) u_2'(1)\lambda_1^2}{\mu_1^2(1 - \rho_1 u_2(1))^2(p - u_2(1))},$$

where

$$u_2(1) = \frac{1}{2\lambda_1}\left(-\sqrt{(\lambda_1 + \mu_1)^2 - 4\lambda_1\mu_1 p} + \lambda_1 + \mu_1\right),$$

$$u_2'(1) = \left.\frac{\partial u_2(v)}{\partial v}\right|_{v=1} = \frac{1}{2\lambda_1}\left(\frac{\lambda_2(\lambda_1 + \mu_1)}{\sqrt{-4\lambda_1\mu_1 p + (\lambda_1 + \mu_1)^2}}\right).$$

Analysing expression (1.61) using Little's theorem, we write: $N_2 = \lambda_2 v_2$, where v_2, as we remember, is the mathematical expectation of the time spent by the request of a second type in the system. On this basis, we have: $v_2 = N_2/\lambda_2 = w_2 + 1/\mu_2$, where the mathematical expectation w_2 of the time spent by the request of the second type in the service station of the system is defined as $w_2 = v_2 - 1/\mu_2$.

Since the requests of the first type have absolute priority over the requests of the second type, the expressions for calculating such stationary characteristics as the mathematical expectation of the number of the requests of the first type in the system N_1, the mathematical expectation of time spent by the request of the first type in the system v_1, and the mathematical expectation of time spent by the request of the first type in the system's service stations w_1 are typical formulas for calculating the corresponding stationary characteristics for the queuing system of type $M|M|1$.

Finally, let's investigate the probability of servicing a request of a second type. We introduce the characteristic parameter s_{ij}—the probability that the selected request of the second type will be served by a system in the service station in which it is preceded by i requests of the first type and j requests of the second type. The process of servicing the selected request of the second type can take place in the following scenarios:

1. Let the selected request of the second type in the system's service station be preceded by i requests of the first type and j requests of the second type: (i, j). Then: with probability $\lambda_1 / (\lambda_1 + \mu_1)$ a request of the first type can come to the system input and take a place before the selected request: $i + 1$; with a probability $\mu_1 / (\lambda_1 + \mu_1)$ can complete the service of one of the requests of the first type, which with a probability p will leave the system without clearing its service station from requests of the second type: $i - 1$.
2. Let the selected request of the second type in the system's service station be preceded only by j requests of the second type: $(0, j)$. Then: with probability $\lambda_1 / (\lambda_1 + \mu_2)$ the input of the system can receive a request of the first type and it takes a place before the selected request: $i + 1$; with probability $\mu_2 / (\lambda_1 + \mu_2)$ can complete the service of one of the j requests of the second type: $j - 1$.
3. Before the selected request of the second type in the system's service station there are no requests: $(0, 0)$. Then: with probability $\lambda_1 / (\lambda_1 + \mu_2)$, the input of the system can receive a request of the first type and it takes a place before the selected request: $i + 1$; with probability $\mu_2 / (\lambda_1 + \mu_2)$ can complete the service of the selected request of the second type.

Scenarios 1 and 2 are represented by the corresponding UML state diagrams in Fig. 1.6.

Analytically describe scenarios 1–3 with the appropriate expressions:

$$s_{i,j} = \frac{\lambda_1}{\mu_1 + \lambda_1} s_{i+1,j} + \frac{\mu_1 p}{\mu_2 + \lambda_1} s_{i-1,j},\ i \geq 1,\ j \geq 1; \tag{1.62}$$

$$s_{0,j} = \frac{\lambda_1}{\mu_2 + \lambda_1} s_{1,j} + \frac{\mu_2}{\mu_2 + \lambda_1} s_{0,j-1},\ j \geq 1; \tag{1.63}$$

$$s_{0,0} = \frac{\lambda_1}{\lambda_1 + \mu_2} s_{1,0} + \frac{\mu_2}{\lambda_1 + \mu_2}. \tag{1.64}$$

We will look for the solution of expression (1.62) in the form

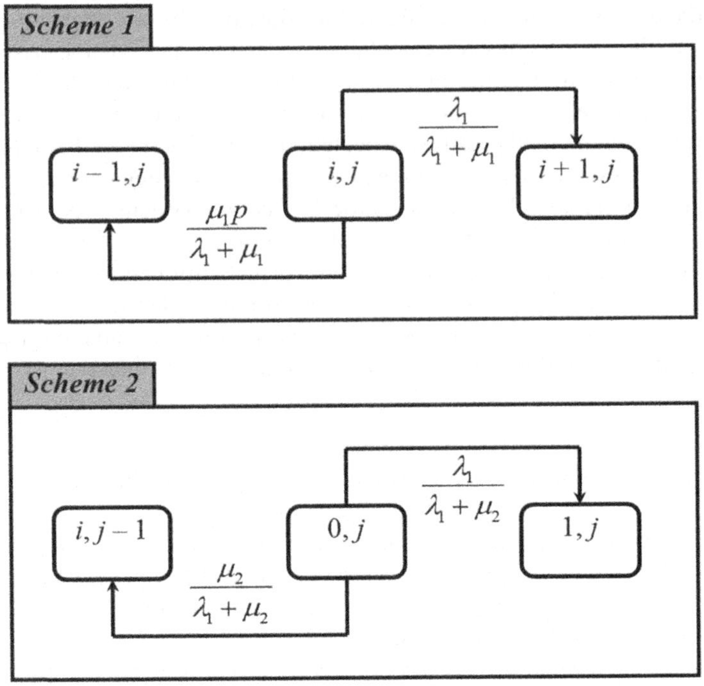

Fig. 1.6 Scenarios of development of the process of servicing the selected request of the second type by the queuing system under study

$$s_{i,j} = \gamma^i s_{0,j},\ i \geq 1,\ j \geq 1. \tag{1.65}$$

Substitute expression (1.65) into expression (1.62) for $i = 1$: $\gamma = \frac{\lambda_1}{\lambda_1+\mu_1}\gamma^2 + \frac{\mu_1 p}{\lambda_1+\mu_1}$. The resulting expression is a quadratic equation concerning γ: $\lambda_1\gamma^2 - (\lambda_1 + \mu_1)\gamma + \mu_1 p = 0$, the solutions of which can be analytically represented as $\gamma_{1,2} = \frac{1}{2\lambda_1}\left(\pm\sqrt{(\lambda_1 + \mu_1)^2 - 4\lambda_1\mu_1 p} + \lambda_1 + \mu_1\right)$. Choosing the root $\gamma_2 \in [0, 1]$ we get

$$s_{i,j} = \gamma_2^i s_{0,j},\ j \geq 0. \tag{1.66}$$

Substitute expression (1.66) into expression (1.63) and express $s_{0,j}$: $s_{0,j} = \frac{\mu_2}{\lambda_1(1-\gamma_2)+\mu_2} s_{0,j-1},\ j \geq 1$. In turn, we substitute expression (1.66) into expression (1.64) for $i = 1, j = 0$, and express $s_{0,0}$: $s_{0,0} = \frac{\mu_2}{\lambda_1(1-\gamma_2)+\mu_2}$. In accordance:

$$s_{0,j} = \delta^{j+1}, j \geq 0, \tag{1.67}$$

where $\delta = \frac{\mu_2}{\lambda_1(1-\gamma_2)+\mu_2}$.

Substitute expression (1.67) into expression (1.66): $s_{i,j} = \gamma_2^i \delta^{j+1}$. Based on the received expression we analytically formalize the required probability of service of request of the second type by expression

$$p^{(serv)} = \sum_{i=0}^{\infty}\sum_{j=0}^{\infty} s_{i,j} p_{i,j} = \delta \sum_{i=0}^{\infty}\sum_{j=0}^{\infty} \delta^j \gamma_2^i p_{i,j} = \delta B(\delta, \gamma_2), \tag{1.68}$$

where the generating function $B(\gamma_2, \delta)$ is defined by expression (1.43).

Let us investigate expression (1.68) under boundary conditions. For $p \to 0$ we have $\gamma_2 \to 0$, $\delta \to \mu_2 / (\lambda_1 + \mu_2)$, and, accordingly, $p^{(serv)} \to \frac{\mu_2}{\lambda_1+\mu_2} B\left(0, \frac{\mu_2}{\lambda_1+\mu_2}\right)$. For $p \to 1$ we have $\gamma_2 \to 1$, $\delta \to 1$, and, accordingly, $p^{(serv)} \to 1$.

We will simulate the dependence of such stationary characteristics of the studied queuing system as—the probability of downtime $p_{0,0}$; mathematical expectation of the number of requests of the second type in the system N_2; the probability of servicing the request of the second type in the system $p^{(serv)}$; from the value of the probability p that the requests of the first type after service leave the system without clearing its service station from the requests of the second type. The established parameters of the system, the functioning of which is simulated, are as follows: $\lambda_1 = 1$, $\lambda_2 = 2$, $\mu_1 = 3$, $\mu_2 = 4$. The results of the experiment are represented by the corresponding curves in Fig. 1.7.

Figure 1.7 shows that with increasing probability p the mathematical expectation of the number of requests of the second type in the studied system N_2 increases and the probability of downtime $p_{0,0}$ decreases. Note the nonlinear quadratic nature of these dependences. This is important because the values of these stationary characteristics

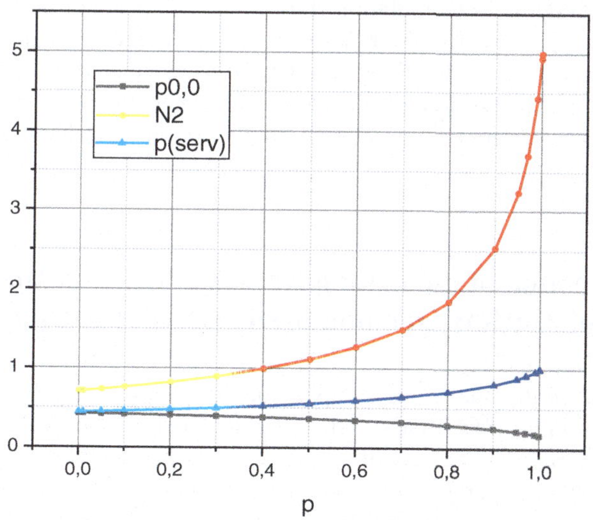

Fig. 1.7 Calculated dependencies of $\{p_{0,0}, N_2, p^{(serv)}\} = f(p; \lambda_1 = 1, \lambda_2 = 2, \mu_1 = 3, \mu_2 = 4)$

for the queuing system of the same type as the studied one, but without the mechanism of the stochastic update, are calculated by the expressions

$$N_2^{(1)} = \frac{\lambda_2(\lambda_1\mu_2 + \mu_1(\mu_1 - \lambda_1))}{(\mu_1\mu_2 - \lambda_1\mu_2 - \lambda_2\mu_1)(\mu_1 - \lambda_1)} = \frac{\rho_2}{-\rho_1 - \rho_2 + 1}\left(\frac{\rho_1\mu_2}{\mu_1(1 - \rho_1)} + 1\right)$$

and

$$p_{0,0}^{(1)} = 1 - \rho_1 - \rho_2, \rho_i = \lambda_i / \mu_i, i = 1, 2,$$

and are equal to 5 and $1/6$, respectively.

Therefore, as $q \rightarrow 0$, the stationary characteristics $p_{0,0}$ and N_2 approach the corresponding values of the classical queueing system without the stochastic update mechanism. This confirms the adequacy of the proposed mathematical apparatus. Additional confirmation of this conclusion is that the probability of downtime for the classical system at $p = 0$ is equal to $p_{0,0}^{(0)} \approx 0, 4269$, which is also consistent with that shown in Fig. 1.7 information. Finally, the fact that the probability $p^{(serv)}$ with increasing value of p goes to one is natural.

Thus, we investigated an instance of DC-like systems that operates under conditions of probable cyber threats (requests of the first type). The implementation of the cyber threat is modelled by a mechanism of stochastic update, the activation of which leads to the loss of all ordinary requests (requests of the second type) in the system's service station. The section formulates expressions for calculating stationary probabilistic characteristics of the studied system, in particular, the probability of downtime, the probability of servicing an ordinary request (in terms of the generating function), the mathematical expectation of the number of ordinary requests in the system. The simulation showed the adequacy of the created mathematical apparatus. The values of stationary probabilistic characteristics of the studied system which functions in the boundary conditions calculated with the help of the proposed mathematical apparatus coincided with the known values calculated for the classical queuing system of a similar type but without the mechanism of stochastic update.

1.3 Investigation of a DC-Like Network with a Sequential Tuple of Markov Queuing Systems with the Update

The applied use of the mathematical apparatus of queuing systems to describe the process of functioning of DC-like systems is limited to determining the control mechanism of service stations. We classify the DC-like system in the first approximation as a queuing network operating in a random environment. More specifically, in this section, we will investigate a tuple formed by sequentially arranged Markov queuing systems with service stations of final capacity and individual mechanisms of generalized update. Such a network simulates the functioning of a base station with a two-factor authentication scheme. Our goal is to define recurrent expressions for calculating the basic stationary characteristics of such a queuing network. Achieving this goal is complicated by the fact that the availability of mechanisms of the update does not allow to classify the studied queuing network as solvable in a multiplicative form. Accordingly, we adapt the matrix-analytical method to formalize the method of calculating the compatible stationary distribution of the number of requests served by the queuing network and the phases of their service by queuing systems within the network. We will also investigate stationary indicators that characterize the situation of loss of incoming requests in the queuing network, in particular, the probability of loss of the received request during its service due to the receipt of new incoming requests to the queuing network.

Therefore, we investigate a network formed by a tuple of sequentially arranged Markov queuing systems of type $M\left|E_{n_i}\right|1|(N_i - 1)$ with service stations of the finite capacity N_i, where n_i is the order of Erlang distribution with the parameter μ_i to describe the stochastic value of the time of servicing a request by the i-th queuing system in the network. We consider the flow of incoming requests to the network Poisson with intensity λ_i. The request is lost if it incomes to the i-th queuing system the capacity of the service station of which is exhausted. In each i-th queuing system the mechanism of the generalized update functions. We describe the essence of this mechanism. If the service station of the i-th system is empty, the serviced request leaves it. If at the end of service of the request by the i-th queuing system in its service station there are n, $1 \leq n \leq N_i$, requests, then: with probability $q_0^{(i)} + Q_n^{(i)}$ the served request leaves the i-th system; with probability $q_j^{(i)}$, $0 < j < n$, the mechanism of update removes from the service station i-th queuing system j requests. Denote by $\left\{q_n^{(i)}, 0 \leq n \leq N_{i-1}\right\}$ the probability of updating in the i-th system $Q_n^{(i)} = q_n^{(i)} + q_{n+1}^{(i)} + \ldots + q_{N_i-1}^{(i)}$; $Q_0^{(i)} = 1, i = 1, 2$. We will assume that requests are serviced and removed from the service station in the order of their receipt. Denote for further use: $1_{(A)}$ is an indicator of the set A; $\vec{e}_{ij}$ is a vector of n_i elements, of which only the j-th element is nonzero; $\mathrm{E}_{km}^{(i)}$ is a matrix of $n_i \times n_i$ elements, of which only the (k, m)-th element is nonzero; $\vec{1}$ is the unit vector; I_n is a unit matrix of dimension $n \times n$.

Considering the stochastic interpretation of the PH-distribution of the type of phase, the functioning of the studied queuing network will be considered as a Markov process $X(t)$, $t \geq 0$, over the space of states

$$X = \{(0, 0)\} \cup \{(i, n, 0), 1 \leq i \leq N_1, 1 \leq n \leq n_1\} \cup$$
$$\cup \{(0, j, m), 1 \leq j \leq N_2, 1 \leq m \leq n_2\} \cup$$
$$\cup \{(i, n, j, m), 1 \leq i \leq N_1, 1 \leq n \leq n_1, 1 \leq j \leq N_2, 1 \leq m \leq n_2\},$$

that is, for any point in time t:

- $X(t) = (0, 0)$—if the service stations of both systems in the network are empty;

- $X(t) = (i, n, 0)((0, j, m))$—if the service station of system 1 is empty, and in the service station of system 2 there are $i(j)$ requests, where $n(m)$ is the number of stages before the end of service of the request in the service station of system 2;
- $X(t) = (i, n, j, m)$—if there are i and j requests on the service stations of systems 1 and 2 and the phases of the requests being serviced are n and m, respectively.

Denote the stationary probability of the state $x \in X$ as $p_x = \lim_{t\to\infty} \mathrm{P}(X(t) = x)$. Based on the logic of formalization of the studied process and assuming the finiteness of the values of the parameters μ_i, n_i, $i = 1, 2$, we consider all states of the process connected. This fact presupposes the existence of corresponding stationary probabilities. Denote the vectors:

$$\begin{aligned}
&\vec{p} = \left(\vec{p}_0, \vec{p}_1, \ldots, \vec{p}_{N_1}\right);\\
&\vec{p}_0 = \left(p_{0,0}, \overline{p_{0,1,1}, p_{0,1,n_2}}, \overline{p_{0,2,1}, p_{0,2,n_2}}, \ldots, \overline{p_{0,N_2,1}, p_{0,N_2,n_2}}\right);\\
&\vec{p}_{i,n} = \left(p_{i,n,0}, \overline{p_{i,n,1,1}, p_{i,n,1,n_2}}, \overline{p_{i,n,2,1}, p_{i,n,2,n_2}}, \ldots, \overline{p_{i,n,N_2,1}, p_{i,n,N_2,n_2}}\right),\\
&\quad 1 \le i \le N_1,\ 1 \le n \le n_1;\ \vec{p}_i = \left(\vec{p}_{i,1}, \ldots, \vec{p}_{i,n_1}\right),\ 1 \le i \le N_1.
\end{aligned}$$

The stationary distribution of the process $X(t)$ is the only solution of the corresponding system of equilibrium equations. In the terminology of processes «generation and death» we will call the level of the process $X(t)$ the number of requests and the phase of service in the service station of system 1, and the phase of the process $X(t)$ the number of requests and the phase of service in the service station of system 2. Then for the studied network, the statement is valid: if the process $X(t)$ is at the level of (i, n), then:

- It is possible to move to a higher level while preserving the current phase (the matrix of intensities of such a transition is denoted by $\mathrm{L}_{(i,n),(k,l)}$).
- It is possible to move to a lower level while preserving the current phase (the matrix of intensities of such a transition is denoted by $\mathrm{N}_{(i,n),(k,l)}$).
- It is possible to move to within the current level with a change of phase (the matrix of intensities of such a transition is denoted by $\mathrm{M}_{(i,n),(k,l)}$).

Since at all levels of the process $X(t)$ the number of phases is the same and equal to $1 + n_2N_2$, then the dimensions of the matrices $\mathrm{L}_{(i,n),(k,l)}$, $\mathrm{N}_{(i,n),(k,l)}$, $\mathrm{M}_{(i,n),(k,l)}$ will be the same, namely, $(1 + n_2N_2) \times (1 + n_2N_2)$. Since the phase of service does not change when the request is received by the system, and when the serviced request is left, the phase of service can be only n_1 (if there is at least one request left in the system), then $\mathrm{L}_{(i,n),(k,l)} = 0$ at $k \ne i + 1$ or $l \ne n$ and $\mathrm{M}_{(i,n),(k,l)} = 0$ at $n \ne 1$ or $l \ne n_1$ or $i \ge k$.

Let us investigate an isolated functioning system to identify the remaining nonzero characteristic matrices. As already mentioned, the process that describes the total number of requests and the phase of service in such a system is Markov. Denote the transition matrix for such a process as $Q^{(i)} = \overset{\frown}{Q}{}^{(i)} + \overset{\smile}{Q}{}^{(i)}$, where

$$\overset{\frown}{Q}^{(i)} = \begin{pmatrix} N_0^{(i)} & L_0^{(i)} & 0 & 0 & \dots & 0 \\ 0 & N_1^{(i)} & L^{(i)} & 0 & \dots & 0 \\ 0 & 0 & N_2^{(i)} & L^{(i)} & \dots & 0 \\ 0 & 0 & 0 & N_3^{(i)} & \dots & 0 \\ \vdots & \vdots & \vdots & \vdots & \ddots & L^{(i)} \\ 0 & 0 & 0 & 0 & \dots & N_{N_i}^{(i)} \end{pmatrix},$$

$$\overset{\smile}{Q}^{(i)} = \begin{pmatrix} 0 & 0 & 0 & 0 & \dots & 0 \\ M_{10}^{(i)} & 0 & 0 & 0 & \dots & 0 \\ 0 & M_{21}^{(i)} & 0 & 0 & \dots & 0 \\ 0 & M_{31}^{(i)} & M_{32}^{(i)} & 0 & \dots & 0 \\ \vdots & \vdots & \vdots & \vdots & \ddots & L^{(i)} \\ 0 & M_{N_i,1}^{(i)} & M_{N_i,2}^{(i)} & M_{N_i,3}^{(i)} & \dots & 0 \end{pmatrix},$$

and matrices $N_0^{(i)}, L_0^{(i)}, L^{(i)}, N_k^{(i)}, M_{kj}^{(i)}$ are defined as

$$N_0^{(i)} = -L_0^{(i)}\vec{1};\ L_0^{(i)} = \lambda_i \vec{e}_{i,n_i};\ L^{(i)} = \lambda_i I_{n_i};$$

$$N_k^{(i)} = -L^{(i)} - \sum_{j=1}^{k-1} M_{kj}^{(i)} E_{n_i,1}^{(i)} + M^{(i)},\ 2 \le k \le N_i - 1;$$

$$M_{kj}^{(i)} = \begin{cases} \mu_i \left(q_0^{(i)} + Q_j^{(i)}\right) E_{1,n_i}^{(i)} \forall k = j+1, \\ \mu_i q_{k-j-1}^{(1)} E_{1,n_i}^{(i)} \forall k \ne j+1, \end{cases}$$

at what

$$N_1^{(i)} = -L^{(i)} - \mu_i E_{11}^{(i)} + M^{(i)};\ N_{N_i}^{(i)} = \sum_{j=1}^{N_i-1} E_{n_i,1}^{(i)} M_{N_i,j}^{(i)} + M^{(i)},$$

where $M^{(i)} = \mu_i \sum\limits_{k=2}^{n_i} \left(E_{k,k-1}^{(i)} - E_{k,k}^{(i)}\right)$.

Based on the analytical expressions just given, we determine the form of nonzero matrices $L_{(i,n),(k,l)}$, $N_{(i,n),(k,l)}$, for the studied network.

Let $i = 0$, i.e. there are no queries in the service station of system 1. From this state, the development of the process $X(t)$ can go either in the direction of change of phase, which will be described by the transition matrix $N_{0,0}$, or in the direction of increasing the level, which will be described by the transition matrix $L_{0,(1,n_1)}$. The receipt of the incoming request transfer the service station of system 1 in the phase of service n_1, but the state of system 2 remains unchanged: $L_{0,(1,n_1)} = \lambda_1 I_{1+n_2N_2}$.

Because the transitions within one layer of the network are identical to the transitions in an isolated system 2, we can write $N_{0,0} = Q^{(2)} - \lambda_i I_{1+n_2N_2}$. Let in the service station of the system 1 is one request and the system is in the phase of service 1. In this case, the development of the process $X(t)$ can occur in the direction of transition to level 0 or 2, which will be regulated by transition matrices $\mathrm{M}_{(1,1),0}$ or $\mathrm{L}_{(1,1),(2,1)}$, respectively, or within the level, (1,1), which will be regulated by transition matrices $\mathrm{N}_{(1,1),(1,1)}$.

Since the transition to level 0, the number of requests in the service station of system 2 will increase by 1 only if there is at least one free space in its service station, the matrix $\mathrm{M}_{(1,1),0}$ of dimension $(1+n_2N_2)\times(1+n_2N_2)$ can be defined as

$$\mathrm{M}_{(1,1),0}=\begin{pmatrix} 0 & \mu_1\vec{e}_{2,n_2} & 0 & \dots & 0\\ 0 & 0 & \mu_1\mathrm{I}_{n_2} & \dots & 0\\ 0 & 0 & & \dots & 0\\ \vdots & \vdots & \vdots & \ddots & \mu_1\mathrm{I}_{n_2}\\ 0 & 0 & 0 & \dots & \mu_1\mathrm{I}_{n_2}\end{pmatrix}. \tag{1.69}$$

Since it was assumed that in the service station of system 1 there is one request, then upon receipt of a new input request, the phase of service in system 1 and the state of system 2 will not change, i.e.: $\mathrm{L}_{(1,1),(2,1)}=\lambda_1\mathrm{I}_{1+n_2N_2}$. Permissible transitions within the level are substantially identical to the transitions defined for an isolated system 2: $\mathrm{N}_{(1,1),(1,1)}=Q^{(2)}-(\lambda_1+\mu_1)\mathrm{I}_{1+n_2N_2}$. For similar reasons, the expressions for calculating the remaining required transition matrices are derived:

$$\mathrm{N}_{(0,0)}=Q^{(2)}-\lambda_1\mathrm{I}_{1+n_2N_2};\ \mathrm{L}_{0,(1,n_1)}=\mathrm{L}_{(i,n)(i+1,n)}=\lambda_1\mathrm{I}_{1+n_2N_2};$$

$$\mathrm{M}_{(i,1),(j,n_1)}=\begin{cases}\left(q_0^{(1)}+Q_j^{(1)}\right)\mathrm{M}_{(1,1),0}\forall i=j+1,\\ q_{i-j-1}^{(1)}\mathrm{M}_{(1,1),0}\forall i\neq j+1,\end{cases} 1\le j\le N_1-1, j+1\le i\le N_1;$$

$$\mathrm{N}_{(1,n),(1,n)}=Q^{(2)}-(\lambda_1+\mu_1)\mathrm{I}_{1+n_2N_2},\ \ 1\le n\le n_1;$$

$$\mathrm{N}_{(i,n+1),(i,n)}=\mu_1\mathrm{I}_{1+n_2N_2},\ \ 1\le n\le n_1-1,\ \ 1\le i\le N_1;$$

$$\mathrm{N}_{(i,n),(i,n)}=Q^{(2)}-\sum_{j=1}^{i-1}\mathrm{diag}\left(\mathrm{M}_{(i,n),(j,n_1)}\vec{1}\right)-\lambda_1\mathrm{I}_{1+n_2N_2}, 1\le n\le n_1, 2\le i\le N_1.$$

We generalize the obtained analytical expressions in the form of a system of equilibrium equations:

$$\vec{p}_{1,1}\mathrm{M}_{(1,1),0}+\vec{p}_0\mathrm{N}_{0,0}=\vec{0}; \tag{1.70}$$

$$\vec{p}_{1,n+1}\mathrm{N}_{(1,n+1),(1,n)}+\vec{p}_{1,n}\mathrm{N}_{(1,n),(1,n)}=\vec{0},\ 1\le n\le n_1-1; \tag{1.71}$$

$$\sum_{i=2}^{N_1}\vec{p}_{i,n_1}\mathrm{M}_{(i,1),(1,n_1)}+\vec{p}_0\mathrm{L}_{0,(1,n_1)}+\vec{p}_{1,n_1}\mathrm{N}_{(1,n_1),(1,n_1)}=\vec{0}; \tag{1.72}$$

$$\vec{p}_{i,n+1}\mathrm{N}_{(1,n+1),(1,n)} + \vec{p}_{i,n}\mathrm{N}_{(1,n),(1,n)} + \vec{p}_{i-1,n}\mathrm{L}_{(i-1,n),(i,n)} = \vec{0},$$
$$1 \le n \le n_1 - 1, 2 \le i \le N_1; \tag{1.73}$$

$$\sum_{j=i+1}^{N_1} \vec{p}_{j,1}\mathrm{M}_{(j,1),(i,n_1)} + \vec{p}_{i-1,n_1}\mathrm{L}_{(i-1,n_1),(i,n_1)} + \vec{p}_{i,n_1}\mathrm{N}_{(1,n_1),(1,n_1)} = \vec{0}, 2 \le i \le N_1 - 1; \tag{1.74}$$

$$\vec{p}_{N_1,n_1}\mathrm{N}_{(N_1,n_1),(N_1,n_1)} + \vec{p}_{N_1-1,n_1}\mathrm{L}_{(N_1-1,n_1),(N_1,n_1)} = \vec{0}; \tag{1.75}$$

$$\sum_{i=1}^{N_1}\sum_{n=1}^{n_1} \vec{p}_{i,n}\vec{1} + \vec{p}_0\vec{1} = 1, \tag{1.76}$$

where expression (1.76) is a normalization condition and $\vec{p}$ is the required vector of state for the studied network.

One of the specialized computational methods [89, 90] can be chosen to solve the system of equilibrium Eqs. (1.70)–(1.75). When choosing a method, we take into account that we are looking for a stationary queue distribution for a system with a service station of a finite capacity with a PH-distribution and a matrix of intensities of type $G|M|1$. At the same time, the strict diagonal predominance inherent in the matrices $\mathrm{N}_{0,0}$ and $\mathrm{N}_{(i,n),(i,n)}$ at any values of i and n ensures their nondegeneracy. This conclusion reduces the order of the system of equilibrium Eqs. (1.70)–(1.75) from $(N_1(n_1-1)+1)\left(1+n_2N_2\right)$ to $N_1\left(1+n_2N_2\right)$, and thus simplifies the process of computation.

We convert Eqs. (1.70), (1.71), (1.75) taking into account the redesignation of $\mathrm{H}_{i,n} = \mathrm{N}_{(i,n+1),(i,n)} \big/ \mathrm{N}_{(i,n),(i,n)}$:

$$\sum_{i=1}^{N_1}\sum_{n=1}^{n_1} \vec{p}_{i,n}\vec{1} + \vec{p}_0\vec{1} = 1, \tag{1.77}$$

$$\mathrm{N}^*_{(i,n),(i,n-1)} \tag{1.78}$$

Convert Eq. (1.73) taking into account the notation $\mathrm{K}_{i,n} = \mathrm{L}_{(i,-1n),(i,n)} \big/ \mathrm{N}_{(i,n),(i,n)}$:

$$\vec{p}_{i,n} = -\mathrm{K}_{i,n}\vec{p}_{i-1,n} - \mathrm{H}_{i,n}\vec{p}_{i,n+1} = \left(\prod_{j=n_1-1}^{n} (-1)^{n_1-n}\mathrm{H}_{i,j}\right)\vec{p}_{i,n_1} - \mathrm{K}_{i,n}\vec{p}_{i-1,n} -$$
$$-\sum_{j=n}^{n_1-1} \mathrm{K}_{i,j}\vec{p}_{i-1,j}\left(\prod_{m=j-1}^{n} (-1)^{n_1-j}\mathrm{H}_{i,j}\right), 1 \le n \le n_1 - 1 \tag{1.79}$$

Finally, the equivalent replacement of the system of equilibrium Eqs. (1.70)–(1.75) is the system of Eqs. (1.72), (1.74), (1.75), in which the vectors of stochastic values

$\vec{p}_0$ and $\vec{p}_{i,1}$ are expressed by Eqs. (1.77)–(1.79). The obtained system of Eqs. (1.72), (1.74), (1.75) is conveniently solved by the method of excluding states [91]. The obtained results should be substituted into Eqs. (1.77)–(1.79) to find as yet unknown stochastic values and to complete the process of calculating the compatible stationary distribution of queues in service stations by the rationing procedure according to expression (1.76).

The found compatible stationary distribution allows to determine such stationary stochastic characteristics of the studied network as the average number of requests $N^{(1)}$ and $N^{(2)}$ in systems 1 and 2, respectively:

$$N^{(1)} = \sum_{i=1}^{N_1} \sum_{n=1}^{n_1} i\vec{p}_{i,n}\vec{1}, \tag{1.80}$$

$$N^{(2)} = \sum_{j=1}^{N_2} \sum_{m=1}^{n_2} p_{0,j,m} j \times \sum_{i=1}^{N_1} \sum_{j=1}^{N_2} \sum_{n=1}^{n_1} \sum_{m=1}^{n_2} p_{i,n,j,m} j. \tag{1.81}$$

The next step is the analytical formalization of the process of calculating the stationary characteristics of the loss of requests in the network under study. We introduce the parameter $\pi^{(i)}$, which characterizes the stationary probability that the request received by the i-th system will not be lost. The nonzero value of the parameter $\pi^{(i)}$ is due to the architecture of the studied network of queuing systems. In particular, the functioning of the mechanisms of the generalized update implemented in the systems in the network may lead to the exclusion of the request received into the system from the service station before its service. Also, the probability of losing a received request is affected by both the number of requests in the service stations in systems in the network that preceded the received request and the number of subsequent incoming requests received by the network until the received request is not serviced by the system 2. Therefore, the incoming request received by the network may be lost:

- while waiting for service in the service station of system 1;
- during the transition to system 2;
- while waiting for service in the service station of system 2.

Denote by $l = (n-1)n_i + m$ the state of the i-th system, in the service station of which there are n requests and the phase of service is equal to m, $1 \le l \le 1 + n_i N_i$. We characterized by the parameter $\left[\mathrm{P}_n(j, i, m)\right]_{kl}$ the conditional probability that the selected request was serviced in system 1 and will enter the service station of system 2, which is in the state l, $1 \le k, l \le 1 + n_2 N_2$. In this case, we take into account that

- in the service station of system 1 there will be n requests;
- the selected request is preceded by $i-1$ requests;
- system 2 is in the state k;
- after the selected request in the service station of system 2 there are j requests;
- the selected request is in system 2 in phase m.

We generalize the following conditional probabilities in the matrices $P_n(j, i, m)$, $j \geq 0, i \geq 1, 0 \leq i+j \leq N_1, 1 \leq m \leq n_1, 0 \leq n \leq N_1 - 1$. The matrices $P_n(j, i, m)$ are related by a recurrent relationship with the matrix of probabilities of transitions of the process $X(t)$. Let us express these relations in the form of an embedded Markov chain generated by the moments of change of states of systems 1 or 2. Let the process $X(t)$ be at level (i, n), i.e. in the service station of system 1 there are i, $1 \leq i \leq N_1$, requests, and the phase of service is equal to n.

Define the transition probability associated with the arrival of a new input request in system 1 and the end of service of the corresponding request in system 2, as an element of the matrix $N^*_{(i,n),(i,n)}$:

$$\left[N^*_{(i,n),(i,n)}\right]_{kl} = \begin{cases} 0 \forall k = l, \\ \dfrac{\left[N_{(i,n),(i,n)}\right]_{kl}}{-\left[N_{(i,n),(i,n)}\right]_{kk}} \forall k \neq l, \end{cases} \quad 1 \leq k, l \leq 1 + n_2 N_2.$$

Define the transition probabilities associated with the change of the phase of service and the receipt of a new input request to the system 1, as elements of the matrices $N^*_{(i,n),(i,n-1)}$ and $L^*_{(i+1,n),(i,n)}$, respectively:

$$\left[N^*_{(i,n),(i,n-1)}\right]_{kl} = \frac{\left[N_{(i,n),(i,n-1)}\right]_{kl}}{-\left[N_{(i,n),(i,n-1)}\right]_{kk}}, \tag{1.82}$$

$$\left[L^*_{(i+1,n),(i,n)}\right]_{kl} = \frac{\left[L_{(i+1,n),(i,n)}\right]_{kl}}{-\left[N_{(i,n),(i,n)}\right]_{kk}}. \tag{1.83}$$

Finally, define the transition probabilities associated with the completion of the service if request in system 1 ($n = 1$) as elements of the matrix $M^*_{(i,1),(j,n_1)}$, $1 \leq j \leq i - 1$:

$$\left[M^*_{(i,1),(i,n_1)}\right]_{kl} = \frac{\left[M_{(i,1),(j,n_1)}\right]_{kl}}{-\left[N_{(i,1),(i,1)}\right]_{kk}}. \tag{1.84}$$

Analytical expressions for determining the elements of matrices $N^*_{0,0}$, $L^*_{0,(1,n_1)}$, $M^*_{(1,1),0}$ are formed for the same reasons that were the basis of expressions (1.82), (1.83), (1.84), respectively.

It is still necessary to describe the event when the selected request after the completion of service in system 1 does not fall into the service of system 2 due to the fullness of the service station of the latter and is lost. According to expression (1.69), such an event is characterized by stochastic elements $\left[M^*_{(i,1),(j,n_1)}\right]_{kl}$ $\left[M^*_{(1,1),0}\right]_{kl}$ of the corresponding transition matrices at $1 \leq k, l \leq 1 + n_2 N_2$. Therefore, when calculating the probability $1 - \pi^{(1)}$ it is necessary to take into account:

$$\left[\mathrm{M}^{*}_{(1,1),0}\right]_{kl} = \left[\mathrm{M}^{*}_{(i,1),(j,n_1)}\right]_{kl} = 0;\ N_2(n_2-1)+1 \le k,l \le n_2N_2+1, \tag{1.85}$$

at $1 \le j \le N_1-1, j+1 \le i \le N_1$.

By analogy, the event when the incoming request received for service in system 1 will not be lost in the service station of this system but will be lost in the service station of system 2, can be taken into account by entering the condition

$$\left[\mathrm{M}^{*}_{(1,1),0}\right]_{kl} = \left[\mathrm{M}^{*}_{(i,1),(j,n_1)}\right]_{kl} = 0;\ 1 \le k,l \le 1+n_2N_2,$$

at $1 \le j \le N_1-1, j+1 \le i \le N_1$.

Therefore, the above considerations, generalized by the formula of total probability, allow us to formalize the recurrent relations for the calculation of matrices $\mathrm{P}_n(j,i,m)$:

$$\begin{gathered}\mathrm{P}_n(j,i,m) = \mathrm{N}^{*}_{(i+j,m),(i+j,m)}\mathrm{P}_n(j,i,m) + \mathrm{N}^{*}_{(i+j,m),(i+j,m-1)}\times\\ \times\mathrm{P}_n(j,i,m-1) + 1_{(j<n)}\mathrm{L}^{*}_{(i+j,m),(i+j+1,m)}\mathrm{P}_n(j+1,i,m),\\ 1\le j\le n, 1\le i\le N_1, 2\le m\le n_1;\end{gathered}$$

$$\begin{gathered}\mathrm{P}_n(j,i,m) = \mathrm{N}^{*}_{(i+j,1),(i+j,1)}\mathrm{P}_n(j,i,1) + 1_{(j<n)}\mathrm{L}^{*}_{(i+j,1),(i+j+1,1)}\mathrm{P}_n(j+1,i,1)\\ + \sum_{k=j+1}^{i+j-1}\mathrm{M}^{*}_{(i+j,1),(k,n_1)}\mathrm{P}_n(j,i-(j+i-k),n_1),\\ 0\le j\le n, 2\le i\le N_1;\end{gathered}$$

$$\mathrm{P}_n(j,i,m) = \begin{cases}\mathrm{N}^{*}_{(j+1,1),(j+1,1)}\mathrm{P}_n(j,1,1) + 1_{(j=n)}\mathrm{M}^{*}_{(j+1,1)(j,n_1)} + \\ +1_{(j<n)}\mathrm{L}^{*}_{(i+j,1),(i+j+1,1)}\mathrm{P}_n(j+1,i,1)\forall 0\le j\le n, n>0,\\ \mathrm{M}^{*}_{(1,1),0} + \mathrm{N}^{*}_{(1,1),(1,1)}\mathrm{P}_0(0,1,1)\forall j=n=0,\end{cases}$$

where $1_{(A)}$ is the indicator of the set.

The matrices $\mathrm{I}-\mathrm{N}^{*}_{(i,m),(i,m)}\forall i, m$ are nondegenerate because they have a strict diagonal predominance. Considering this fact, the matrices $\mathrm{P}_0(0,i,m)$ are successively calculated starting with $\mathrm{P}_0(0,1,1)$ using the expressions:

$$\mathrm{P}_0(0,1,1) = \frac{\mathrm{M}^{*}_{(1,1),0}}{\mathrm{I}_{1+n_2N_2} - \mathrm{N}^{*}_{(1,1),(1,1)}}; \tag{1.86}$$

$$\mathrm{P}_0(0,i,1) = \frac{\sum_{k=1}^{i-1}\mathrm{P}_0(0,k,n_1)\mathrm{M}^{*}_{(i,1),(k,n_1)}}{\mathrm{I}_{1+n_2N_2} - \mathrm{N}^{*}_{(i,1),(i,1)}},\ 2\le i\le N_1; \tag{1.87}$$

$$\mathrm{P}_0(0,i,m) = \frac{\mathrm{P}_0(0,i,m-1)\mathrm{N}^{*}_{(i,m),(i,m-1)}}{\mathrm{I}_1 - \mathrm{N}^{*}_{(i,m),(i,m)}},\ 1\le i\le N_1,\ 2\le m\le n_1. \tag{1.88}$$

The same considerations that allowed us to obtain expressions (1.88)–(1.86) apply to the conversion of expressions for the calculation of matrices $\mathrm{P}_n(j, i, m)$, $1 \le n \le N_1 - 1$:

$$\mathrm{P}_n(j, 1, 1) = \frac{1_{(j<n)}\mathrm{L}^*_{(j+i,1),(j+i+1,1)}\mathrm{P}_n(j+1, i, 1) + 1_{(j=n)}\mathrm{M}^*_{(j+1,1),(j,n_1)}}{\mathrm{I}_{1+n_2N_2} - \mathrm{N}^*_{(j+1,1),(j+1,1)}},\ 1 \le j \le n, \tag{1.89}$$

$$\mathrm{P}_n(j, i, 1) = \frac{\sum\limits_{k=j+1}^{j+i-1} \mathrm{M}^*_{(j+i,1),(k,n_1)}\mathrm{P}_n(j, k-j, n_1) + 1_{(j<n)}\mathrm{L}^*_{(j+i,1),(j+i+1,1)}\mathrm{P}_n(j+1, i, 1)}{\mathrm{I}_{1+n_2N_2} - \mathrm{N}^*_{(j+i,1),(j+i,1)}}, \tag{1.90}$$

$0 \le j \le n,\ 2 \le i \le N_1,$

$$\mathrm{P}_n(j, i, m) = \frac{\mathrm{N}^*_{(j+i,m),(j+i,m-1)}\mathrm{P}_n(j, i, m-1) + 1_{(j<n)}\mathrm{L}^*_{(j+i,m),(j+i+1,m)}\mathrm{P}_n(j+1, i, m)}{\mathrm{I}_{1+n_2N_2} - \mathrm{N}^*_{(j+i,m),(j+i,m)}}, \tag{1.91}$$

$0 \le j \le n,\ 2 \le i \le N_1,\ 2 \le m \le n_1.$

We generalize the order of application of expressions (1.88)–(1.89) for calculation of matrices $\mathrm{P}_n(j, i, m)$ in the form of the UML activity diagram shown in Fig. 1.8.

Based on the calculated matrices $\mathrm{P}_n(j, i, m)$, we formulate an expression that characterizes the probability of successful passage of the request through the system 1:

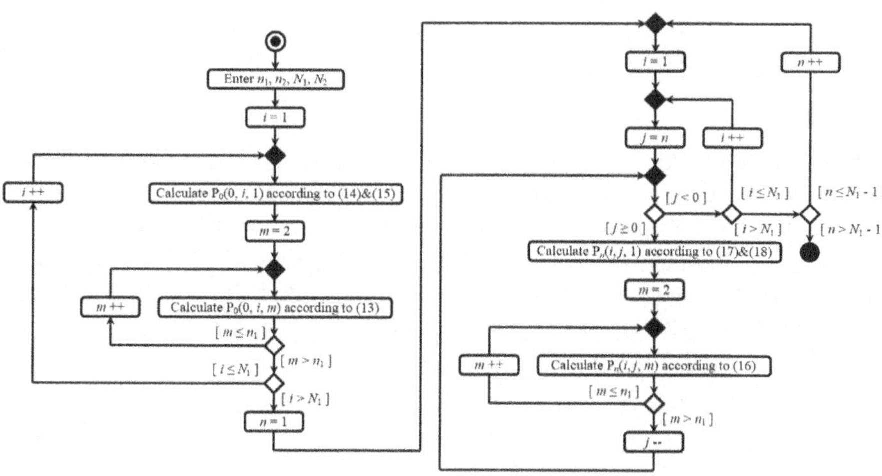

Fig. 1.8 UML activity diagrams of the process of calculation of matrices $\mathrm{P}_n(j, i, m)$

$$\sum_{m=1}^{N_1-1}\sum_{j=1}^{n_1}\sum_{n=0}^{N_1-1} \mathrm{P}_n(0,m,j)\vec{p}_{mj}\vec{1} + \sum_{n=0}^{N_1-1} \mathrm{P}_n(0,1,n_1)\vec{p}_0\vec{1}. \tag{1.92}$$

If expression (1.92) is calculated taking into account condition (1.84), then the obtained stochastic value will characterize the probability that the request will successfully pass system 1 and get to the service station of system 2. If expression (1.92) is calculated taking into account condition (1.85), the obtained stochastic value will characterize the probability that the request will successfully pass system 1, but will not get to the service station of system 2 and will be lost. Summarizing these two stochastic values, we obtain the value of $1-\pi_1^{(1)}$, which characterizes the probability that the request will not be lost in system 1.

Define the vectors $\vec{p}_0^*$, $\vec{p}_{i,n}^*$, $1 \le i \le N_1 - 1$, $1 \le n \le n_1$, arbitrary, l-th, $1 \le l \le 1 + n_2N_2$, the element $\left[\vec{p}_{i,n}^*\right]_l\left(\left[\vec{p}_0^*\right]_l\right)$ of which characterizes the unconditional probability that the request will get into the service station of system 2 and at the time of its receipt:

- system 2 will be in the state l;
- in the service station of system 1 there will be i requests at the stage of service n (if $i \ge 1$):

$$\vec{p}_0^* = \sum_{m=1}^{N_1-1}\sum_{j=1}^{n_1} \vec{p}_{ij}\mathrm{P}_0(0,m+1,j) + \vec{p}_0\mathrm{P}_0(0,1,n_1);$$

$$\vec{p}_{i,n}^* = \begin{cases} \sum_{m=1}^{N_1-1}\sum_{j=1}^{n_1} \vec{p}_{mj}\mathrm{P}_i(0,m+1,j) + \vec{p}_0\mathrm{P}_i(0,1,n_1)\forall n = n_1, \\ 0\forall 1 \le n \le n_1 - 1,\ 1 \le i \le N_1 - 1, \end{cases}$$

$$\vec{p}_{N_1,n}^* = \vec{0},\ 1 \le n \le n_1.$$

After passing system 1, the request enters the service station of system 2, where it can be preceded by no more than N_2 requests. Each state of system 2 corresponds to one of the $1 + n_1N_1$ states of system 1. Define the vectors $\vec{p}_0^+$, $\vec{p}_{j,m}^+$, $1 \le j \le N_2$, $1 \le m \le n_2$, arbitrary, l-th, $1 \le l \le 1 + n_2N_2$, element $\left[\vec{p}_{i,n}^+\right]_l\left(\left[\vec{p}_0^+\right]_l\right)$ of which characterizes the unconditional probability that the request will get into the service station of system 2 and at the time of its receipt:

- system 1 will be in the state l, $1 \le l \le 1 + n_1N_1$;
- in the service station of system 2 there will be j requests at the stage of service m (if $j \ge 1$):

$$\vec{p}_0^+ = \left(\left[\vec{p}_0^*\right]_1, \overline{\left[\vec{p}_{1,1}^*\right]_1, \left[\vec{p}_{1,n_1}^*\right]_1}, \ldots, \overline{\left[\vec{p}_{N_1-1,1}^*\right]_1, \left[\vec{p}_{N_1-1,n_1}^*\right]_1}, \overline{\left[\vec{p}_{N_1,1}^*\right]_1, \left[\vec{p}_{N_1,n_1}^*\right]_1}\right);$$

$$\vec{p}^{\,+}_{j,m} = \Big(\big[\vec{p}^{\,*}_0\big]_{1+m+(j-1)n_2}, \overline{\big[\vec{p}^{\,*}_{1,1}\big]_{1+m+(j-1)n_2}, \big[\vec{p}^{\,*}_{1,n_1}\big]_{1+m+(j-1)n_2}}, \ldots,$$
$$\overline{\big[\vec{p}^{\,*}_{N_1-1,1}\big]_{1+m+(j-1)n_2}, \big[\vec{p}^{\,*}_{N_1-1,n_1}\big]_{1+m+(j-1)n_2}}, \overline{\big[\vec{p}^{\,*}_{N_1,1}\big]_{1+m+(j-1)n_2}, \big[\vec{p}^{\,*}_{N_1,n_1}\big]_{1+m+(j-1)n_2}}\Big).$$

To obtain an analytical description of the event when the request received in system 2 will not be lost, we can repeat the reasoning embodied in expression (1.92). It is necessary to calculate matrices $\tilde{\mathrm{P}}_n(j, i, m), j \geq 0, i \geq 1, 0 \leq i+j \leq N_2, 1 \leq m \leq n_2$, $0 \leq n \leq N_2 - 1$, in which the element $1 \leq k, l \leq 1 + n_1 N_1$ characterizes the conditional probability of the event when received in system 2 the selected request will be serviced and at the end of its service the system 1 will be in the state l and in the service station of system 2 will remain n requests. Conditions for the implementation of such an event are.

- the selected request in the service station of system 2 is preceded by $i-1$ requests;
- the phase of service of the selected request in system 2 is equal to m;
- after the selected request in the pipeline of system 2 requests;
- system 1 is in the state k.

Since the service of the selected request in system 2 is preceded by its service in system 1, it is necessary to modify the matrices of transition intensities, turning them into the form:

$$\tilde{\mathrm{N}}_{(0,0)} = \overset{\frown}{Q}^{(1)} - \lambda_2 \mathrm{I}_{n_2 N_1+1} - \mathrm{diag}\Big(\overset{\smile}{Q}^{(1)}\vec{1}\Big); \tilde{\mathrm{L}}_{(i,n),(j,m)} = \lambda_2 \mathrm{I}_{n_1 N_1+1} + \overset{\smile}{Q}^{(1)};$$

$$\tilde{\mathrm{M}}_{(i,1),(j,n_1)} = \begin{cases} \Big(q_0^{(2)} + Q_j^{(12)}\Big)\tilde{\mathrm{M}}_{(1,1),0} \forall i = j+1, \\ q_{i-j-1}^{(2)} \mathrm{M}_{(1,1),0} \forall i \neq j+1, \end{cases} \quad 1 \leq j \leq N_2 - 1, j+1 \leq i \leq N_2;$$

$$\tilde{\mathrm{N}}_{(1,n+1),(1,n)} = \overset{\frown}{Q}^{(1)} - \mathrm{diag}\Big(\overset{\smile}{Q}^{(1)}\vec{1}\Big) - (\lambda_2 + \mu_2)\mathrm{I}_{n_1 N_1+1}, 1 \leq n \leq n_2;$$

$$\tilde{\mathrm{N}}_{(i,n+1),(i,n)} = \mathrm{I}_{n_1 N_1+1}\mu_2, 1 \leq n \leq n_2 - 1, 1 \leq i \leq N_2;$$

$$\tilde{\mathrm{N}}_{(i,1),(i,1)} = \overset{\frown}{Q}^{(1)} - \sum_{j=1}^{i-1} \mathrm{diag}\Big(\tilde{\mathrm{M}}_{(i,1),(j,n_1)}\vec{1}\Big) - \mathrm{diag}\Big(\overset{\smile}{Q}^{(1)}\vec{1}\Big) - \lambda_2 \mathrm{I}_{n_1 N_1+1},$$
$$1 \leq n \leq n_2, 2 \leq i \leq N_2.$$

We use these modified matrices of transition intensities to calculate the matrices of transitions $\tilde{\mathrm{P}}_n(j, i, m)$ as a result of the implementation of the computational procedure represented by the UML activity diagram in Fig. 1.8, replacing in the letter n_1 by n_2 and N_1 by N_2.

Based on the calculated matrices $\tilde{\mathrm{P}}_n(j, i, m)$, we formulate an expression that characterizes the total probability of request loss in system 1:

$$1 - \pi^{(1)} = \sum_{j=1}^{N_2-1} \sum_{m=1}^{n_2} \sum_{n=0}^{N_2-1} \mathrm{P}_n(0, j+1, m) \vec{p}_{j,m}^{\,+} \vec{1} + \vec{p}_0^{\,+} \vec{1}, \tag{1.93}$$

where $\vec{p}_0^{\,+}\vec{1}$ is the probability that the served in system 1 request came to the empty service station of system 2.

If you change the vectors $\vec{p}_0^{\,+}\vec{1}$ and $\vec{p}_{j,m}^{\,+}$ in expression (1.93) concerning the PASTA-characteristic of the Poisson flow:

$$\vec{p}_0^{\,+} = \left(p_{0,0}, \overline{p_{1,1,0}, p_{1,n_1,0}}, \overline{p_{2,1,0}, p_{2,n_1,0}}, \ldots, \overline{p_{N_1,1,0}, p_{N_1,n_1,0}}\right);$$

$$\vec{p}_{j,m}^{\,+} = \left(p_{0,,j,m}, \overline{p_{1,j,m}, p_{1,n_1,j,m}}, \overline{p_{2,1,j,m}, p_{2,n_1,j,m}}, \ldots, \overline{p_{N_1,1,j,m}, p_{N_1,n_1,j,m}}\right),$$
$$1 \le j \le N_2,\ 1 \le m \le n_2,$$

then this expression can be used to calculate the stochastic value of $1 - \pi^{(2)}$, which characterizes the total probability of loss of the request in system 2.

In summary, the proposed mathematical apparatus for formalizing systems of equilibrium equations that characterize a network tuple of two sequentially connected queuing systems with generalized mechanisms of update can be scaled to describe network tuple of *n* sequentially connected corresponding queuing systems. However, as the number of described systems in the network tuple increases, the dimension of the corresponding characteristic matrices increases rapidly. The calculation of the time of the request stay in the network, which is modelled by a Markov chain with a finite state space, can be done in terms of Laplace–Stieltjes transformation, because the values of time intervals between transitions of a nested Markov network are described by different but exponential distributions. But, considering the computational complexity of expressions (1.88)–(1.89), it becomes obvious that even for a network consisting of a tuple of two systems, the practical calculation of time characteristics should be implemented using the Little theorem. Under such conditions, further research should be directed towards finding more efficient computational methods for solving the corresponding systems of equilibrium equations.

1.4 Evaluation of Functional Safety Indicators in Cloud-Based DC-Like Systems

The socially proclaimed transition to the doctrine of Industry 4.0 led to the mas introduction of cyber-physical devices in the industrial and social spheres of human activity. Under such conditions, the issue of integration and processing of information in such distributed information systems became acute. The answer was the creation of distributed cloud computing technology. However, the well-known advantages of this technology do not eliminate its inherent problems, which are now being actively studied in the taxonomy of dependability in general [92–95], and its attributes such as reliability, availability and functional safety in particular. The source of these problems is the sensitivity of cloud computing technology to failures in active user and system information processes, which lead to reduced performance, breakdowns or shutdown of cloud information systems.

Functional safety provisions provide for the presence in the cloud information system of appropriate security mechanisms, the critical parameter of which is the time required for the correct shutdown of the system after disabling the flow of incoming requests. We will connect this parameter with the concept of the maximum residual time of service, which means the maximum of the set of values of the residual time of service for all involved service devices of the queuing system at the time T. As a model of a DC-like cloud information system, we take a queuing system of type $M^{(2)}\big|G_2|\infty$, to which the Poisson flow of requests with intensity λ is directed. As is typical for DC-like systems, upon receipt, the incoming request is divided into two equally structured subrequests, which are sent for service to the respective subsystems. Each of these subsystems does not contain a service station, but includes an infinite number of service devices and is characterized by a corresponding distribution law with a distribution function $B_i(x)$, $i = 1, 2$. Accordingly, we will investigate the two-dimensional distribution of the values of the maximum residual time of service of subrequests in both subsystems. The difference of such a system from the classical DC-system is that in the latter to the structure of each similar subsystem includes a service station of unlimited capacity, generalized by one service device. However, in articles [96–98] systems with a structure similar to the one studied in this section were investigated. These studies present analytical expressions for the generating functions of the probability distribution of the Markov chain, which characterizes the number of requests in each subsystem, but such a parameter as the maximum residual time of service was ignored.

Let us investigate the functioning of a system of the type $M^{(2)}\big|G_2|\infty$ under conditions when the intensity of the input Poisson flow of requests does not depend on time. Let us estimate the distribution function $G_T(x.y)$ of the two-dimensional stochastic value of the maximum of values of the residual time of service of subrequests in the subsystems of such a system at the time T ($T = \infty$ corresponds to the stationary mode).

If $\lambda = const$, $0 < T < \infty$ then

$$G_T(x, y) = \exp\left(-\lambda \int_0^T (1 - B_1(t + x)B_2(t + y))dt\right), \tag{1.94}$$

$$G_\infty(x, y) = \exp\left(-\lambda \int_0^\infty (1 - B_1(t + x)B_2(t + y))dt\right), \tag{1.95}$$

where x and y are independent stochastic values that characterize the time of service of subrequests of one request in subsystems 1 and 2, respectively.

Let us represent expression (1.94) as follows:

$$G_T(x,y) = \exp\left(-\lambda\int_0^T \overline{B}_1(t+x)dt - \lambda\int_0^T \overline{B}_2(t+y)dt + \right.$$
$$\left. +\lambda\int_0^T \overline{B}_1(t+x)\overline{B}_2(t+y)dt\right) = G_{1,T}(x)G_{2,T}(y)G_T(x,y), \tag{1.96}$$

where

$$\overline{B}_1(t+x) = 1 - B_1(t+x), \overline{B}_2(t+y) = 1 - B_{21}(t+y);$$
$$G_{1,T}(x) = \exp\left(\lambda\int_0^T \overline{B}_1(t+x)dt\right), G_{2,T}(y) = \exp\left(\lambda\int_0^T \overline{B}_2(t+y)dt\right)$$

are partial functions of distribution of values of the maximum residual time of service of subrequests in the first and second subsystems, respectively, and

$$G_T(x,y) = \exp\left(\lambda\int_0^T \overline{B}_1(t+x)\overline{B}_2(t+y)dt\right)$$

is a multiplier that characterizes their dependence.

We introduce the distribution function

$$F_{1,T}(x) = 1 - \frac{1}{\mu_{1,T}}\int_0^T \overline{B}_1(t+x)dt, F_{2,T}(x) = 1 - \frac{1}{\mu_{2,T}}\int_0^T \overline{B}_2(t+y)dt, \tag{1.97}$$

where $\mu_{1,T} = \int_0^T \overline{B}_1(t)dt$, $\mu_{2,T} = \int_0^T \overline{B}_2(t)dt$.

The dependence $G_T(x,y) = D_T(x.y)\exp\left(-\lambda\mu_{1,T}\overline{F}_{1,T}(x)\right)\exp\left(-\lambda\mu_{2,T}\overline{F}_{2,T}(y)\right)$ will be investigated in the paradigm of copula theory [99]. As is known, the copula C is a function of a multidimensional distribution defined on $[0,1]^d$, $d \geq 2$, all partial distributions of which on [0, 1] are uniform. According to Sklyar's theorem, an arbitrary function of such a multidimensional distribution can be represented as $F(x_1,\ldots,x_d) = C(F_1(x_1),\ldots,F_d(x_d))$, where F_i, $i = \overline{1,d}$ are functions of partial distributions. Therefore, each multidimensional distribution can be characterized by a corresponding copula, and if the partial distributions are continuous, the corresponding copula is unique. If we characterize stochastic vectors (X,Y) (with continuous distribution functions F_X, F_Y of their components) by two-dimensional copulas $C(u,v)$, then for a compatible distribution function of X and Y there is a single representation: $F_{X.Y}(x.y) = C(F_X(x), F_Y(y))$.

It is characteristic of the studied system that the distribution functions $G_{1,T}(x)$, $G_{2,T}(y)$ at point $x = y = 0$ show a jump to a certain positive value as an indicator of the presence of both subsystems in the standby state. Therefore, a copula

$$C_T(u, v) = uv \times D_T\left(1/G_{1,T}(u), 1/G_{2,T}(v)\right) \tag{1.98}$$

for expression (1.96) defined only if conditions $u \geq G_{1,T}(x)$, $v \geq G_{2,T}(y)$ are satisfied.

The dependence in expression (1.98) is conveniently characterized by a certain finite numerical value, for example, the Blomqvist (Pearson, Kendell, Spearman, etc.) correlation coefficient. This coefficient $\beta_{X.Y}$ for stochastic quantities X and Y is calculated by the expression $\beta_{X.Y} = \mathrm{E} \times \mathrm{sign}(X - X_m)(Y - Y_m)$, where E is the unit matrix of the corresponding dimension and X_m, Y_m are the medians of the quantities X and Y, respectively.

The coefficient $\beta_{X.Y}$ can be determined through the copula as

$$\beta_C = 4C\left(1/2, 1/2\right). \tag{1.99}$$

Note that for the system under study, the medians X_m and Y_m are positive only if the conditions $G_{1,T}(x) \leq 1/2$ and $G_{2,T}(y) \leq 1/2$ are satisfied for the stochastic quantities X and Y, respectively.

Assume that the stochastic quantities X and Y, which characterize the time of service of subrequests in the first and second subsystems, respectively, are distributed according to the exponential distribution law:

$$B_1(x) = 1 - \exp(-\beta_1 x), B_2(y) = 1 - \exp(-\beta_2 y), \; x, y \geq 0, \; \beta_1, \beta_2 > 0. \tag{1.100}$$

Then expressions (1.94), (1.95) are represented as

$$\begin{gathered}
G_T(x, y) = \exp\Bigg(-\lambda\bigg(\frac{1}{\beta_1}(1 - \exp(-\beta_1 T)) + \\
+\frac{1}{\beta_2}\exp(-\beta_2 y)(1 - \exp(-\beta_2 T))\exp(-\beta_1 x) - \\
-\frac{1}{\beta_1 + \beta_2}(1 - \exp(-T(\beta_1 + \beta_2)))\exp(-\beta_1 x + \beta_2 y)\bigg)\Bigg), \\
x, y \geq 0, 0 \leq T \leq \infty;
\end{gathered} \tag{1.101}$$

$$\begin{gathered}
G_\infty(x, y) = \exp\Bigg(-\lambda\bigg(-\frac{1}{\beta_1 + \beta_2}\exp(\beta_1 x + \beta_2 y) + \\
+\frac{1}{\beta_1}\exp(-\beta_1 x) + \frac{1}{\beta_2}\exp(-\beta_2 y)\bigg)\Bigg), \\
x, y \geq 0.
\end{gathered}$$

The partial functions of the distribution of the values of the maximum residual time of service in the first and second subsystems $G_{1,T}(x)$, $G_{2,T}(y)$ based on expression (1.101) are represented as

$$G_{1,T}(x) = \exp\left(-\frac{\lambda}{\beta_1}\exp(-\beta_1 x)(1-\exp(-\beta_1 T))\right),$$
$$G_{2,T}(y) = \exp\left(-\frac{\lambda}{\beta_2}\exp(-\beta_2 y)(1-\exp(-\beta_2 T))\right). \quad (1.102)$$

We define the inverse functions to expressions (1.102) as

$$\frac{1}{G_{1,T}(u)} = -\frac{1}{\beta_1}\ln\left(-\frac{\beta_1 \ln u}{\lambda(1-\exp(-\beta_1 T))}\right),$$
$$\frac{1}{G_{2,T}(v)} = -\frac{1}{\beta_2}\ln\left(-\frac{\beta_2 \ln v}{\lambda(1-\exp(-\beta_2 T))}\right) \quad (1.103)$$

Taking into account expressions (1.101) and (1.103), we present expression (1.98) to define the copulas $C_T(u, v)$, $C_\infty(u, v)$ as

$$C_T(u, v) = G_T\left(\frac{1}{G_1(u)}, \frac{1}{G_2(v)}\right) =$$
$$= uv\exp\left(\frac{1-\exp(-\beta_1 T-\beta_2 T)}{(1-\exp(-\beta_1 T))(1-\exp(-\beta_2 T))} \times \frac{\beta_1\beta_2 \ln u \ln v}{\lambda(\beta_1+\beta_2)}\right), \quad (1.104)$$
$$C_\infty(u, v) = uv \times \exp\left(\frac{\beta_1\beta_2 \ln u \ln v}{\lambda(\beta_1+\beta_2)}\right).$$

The correlation coefficient (1.99) for the copula $C_\infty(u, v)$ is defined as

$$\beta_C = 4C_\infty\left(\frac{1}{2}, \frac{1}{2}\right) - 1 = \exp\left(\ln^2\frac{1}{2} \times \frac{\beta_1\beta_2}{\lambda(\beta_1+\beta_2)}\right) - 1 =$$
$$= 2 \wedge \left(\frac{\beta_1\beta_2 \ln 2}{\lambda(\beta_1+\beta_2)}\right) - 1. \quad (1.105)$$

Graphical interpretation of expression (1.104) in parametric space (u, v) and such values of fixed parameters as $\lambda = 2$, $\beta_1 = \beta_2 = 1$ is presented in Fig. 1.9.

Assume that the stochastic values of X and Y, which characterize the time of service of subrequests in the first and second subsystems, respectively, are distributed according to the hyperexponential distribution law:

$$B_1(x) = 1 - \sum_{i=1}^{n} c_{1i}\exp(-\beta_{1i}x),\ x \geq 0,\ \beta_{1,i} > 0,\ c_{1i} \geq 0,\ \sum_{i=1}^{n} c_{1i} = 1,$$

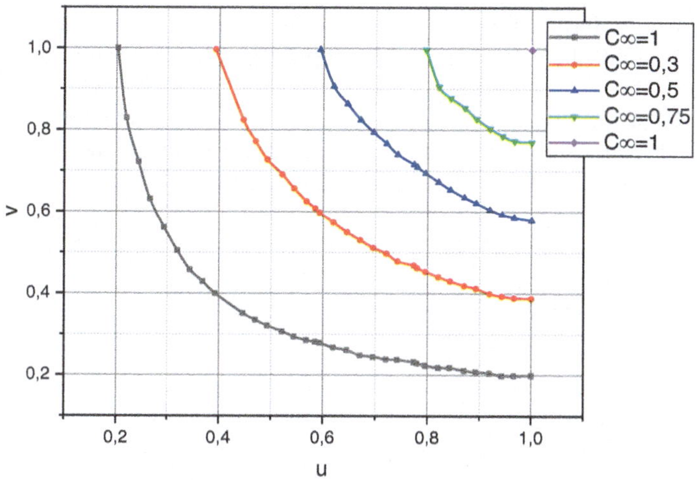

Fig. 1.9 2D representation of the dependence $C_T = f(u, v; \lambda = 2, \beta_1 = \beta_2 = 1)$ calculated by expression (1.104)

$$B_2(x) = 1 - \sum_{i=1}^{n} c_{2i} \exp(-\beta_{2i} y),\ y \geq 0,\ \beta_{2,i} > 0,\ c_{2i} \geq 0,\ \sum_{i=1}^{n} c_{2i} = 1.$$

Then expressions (1.94), (1.95) are represented as

$$G_T(x, y) = \exp\Bigg(-\lambda\Bigg(\sum_{i=1}^{n} \frac{c_{1i}}{\beta_{1i}}(1 - \exp(-\beta_{1i}T)\exp(-\beta_{1i}x)) + \\ + \sum_{i=1}^{n} \frac{c_{2i}}{\beta_{2i}}(1 - \exp(-\beta_{2i}T)\exp(-\beta_{2i}y)) - \\ - \sum_{i=1}^{n}\sum_{i=1}^{n} \frac{c_{1i}c_{2i}}{\beta_{1i} + \beta_{2i}}(1 - \exp(-T(\beta_{1i} + \beta_{2i})\exp(-\beta_{1i}x + \beta_{2i}y)))\Bigg)\Bigg), \\ x, y \geq 0, 0 \leq T \leq \infty;$$

$$G_\infty(x, y) = \exp\Bigg(-\lambda\Bigg(-\sum_{i=1}^{n}\sum_{i=1}^{n} \frac{c_{1i}c_{2i}}{\beta_{1i} + \beta_{2i}} \exp(\beta_{1i}x + \beta_{2i}y) + \\ + \sum_{i=1}^{n} \frac{c_{1i}}{\beta_{1i}} \exp(-\beta_{1i}x) + \sum_{i=1}^{n} \frac{c_{2i}}{\beta_{2i}} \exp(-\beta_{2i}x)\Bigg)\Bigg), \\ x, y \geq 0.$$

Unfortunately, further transformations of the obtained expressions to formalize the expressions for the calculation of the copula and the correlation coefficient are impossible.

Assume that the stochastic values of X and Y, which characterize the time of service of subrequests in the first and second subsystems, respectively, are distributed according to the Pareto distribution law with parameter $\alpha = 2$:

$$B_1(x) = 1 - (x+1)^{-2},\ B_2(y) = 1 - (y+1)^{-2},\ x, y \geq 0.$$

Then expressions (1.94), (1.95) are represented as

$$\begin{aligned} G_T(x,y) = &\exp\Bigg(-\lambda\Bigg(\frac{(x-y+1)(x-y-1)}{(x-y)^2} + \\ &+\Bigg(-\frac{2}{(x-y)^3}\ln\frac{(y+1)(x+T+1)}{(x+1)(y+T+1)} + \\ &+\frac{T}{(x+1)(x+T+1)} + \frac{T}{(y+1)(y+T+1)}\Bigg) - \Bigg), \\ &x, y \geq 0, 0 < T \leq \infty, x \neq y; \end{aligned} \tag{1.106}$$

$$\begin{aligned} G_\infty(x,y) = \exp\Bigg(-\lambda\Bigg(\Bigg(\frac{1}{x+1} + \frac{1}{y+1}\Bigg) - \frac{2}{(x-y)^3}\ln\frac{y+1}{x+1}\Bigg)\times \\ \times\frac{(x-y+1)(x-y-1)}{(x-y)^2}\Bigg), \\ x, y \geq 0, x \neq y. \end{aligned}$$

The partial functions of the distribution of the values of the maximum residual time of service in the first and second subsystems $G_{1,T}(x)$, $G_{2,T}(y)$ based on expression (1.101) are represented as

$$\begin{aligned} G_{1,T}(x) &= \exp\Bigg(-\lambda\frac{T}{(x+1)(x+T+1)}\Bigg), \\ G_{2,T}(x) &= \exp\Bigg(-\lambda\frac{T}{(y+1)(y+T+1)}\Bigg), \end{aligned}$$

$$G_{1,\infty}(x) = \exp\Bigg(-\lambda\frac{1}{(x+1)}\Bigg),\ G_{2,\infty}(y) = \exp\Bigg(-\lambda\frac{1}{(y+1)}\Bigg). \tag{1.107}$$

We define the inverse functions to expressions (1.107) as

$$\frac{1}{G_{1,\infty}(u)} = -\lambda\frac{1}{\ln u} - 1,\ \frac{1}{G_{2,\infty}(v)} = -\lambda\frac{1}{\ln v} - 1. \tag{1.108}$$

Taking into account expressions (1.106) and (1.108), expression (1.98) for the definition of the copula $C_\infty(u, v)$ is represented as

$$C_\infty(u, v) = uv \exp\left(\frac{\ln^2 u \ln^2 v\left(2 \ln u \ln v \ln\left(\frac{\ln u}{\ln v}\right) - \ln(uv) \ln\left(\frac{u}{v}\right)\right)}{\lambda^2(\ln u - \ln v)^3}\right), u \neq v. \quad (1.109)$$

Expressions (1.107) and (1.109) at $x = y$ taking the form:

$$G_T(x, y = x) = \exp\left(-\lambda\left(\frac{1}{3(x + T + 1)^3} - \frac{2}{x + T + 1} - \frac{1}{3(x + 1)^3} + \frac{2}{x + 1}\right)\right),$$
$$x \geq 0, 0 < T \leq \infty,$$

$$G_\infty(x, y = x) = \exp\left(-\lambda\left(\frac{2}{x + 1} - \frac{1}{3(x + 1)^3}\right)\right), x \geq 0,$$

$$C_\infty(u, v = u) = u^2 \exp\left(-\frac{\ln^3 u}{3\lambda^2}\right).$$

The correlation coefficient (1.99) for copula $C_\infty(u, v)$ is significant as

$$\beta_C = 4C_\infty\left(\frac{1}{2}, \frac{1}{2}\right) - 1 = \exp\left(-\frac{\ln^3 1/2}{3\lambda^2}\right) - 1 = 2^\wedge\left(\frac{\ln^2 2}{3\lambda^2}\right) - 1. \quad (1.110)$$

Graphical interpretation of expression (1.109) in the parametric space (u, v) and such values of fixed parameters $\lambda = 2, \alpha = 2$ is presented in Fig. 1.10.

Assume that the stochastic values of X and Y, which characterize the time of service of subrequests in the first and second subsystems, respectively, are distributed according to the Pareto distribution law, but with the parameter $\alpha = 3$. In accordance:

$$B_1(x) = 1 - (x + 1)^{-3},\ B_2(y) = 1 - (y + 1)^{-3},\ x, y \geq 0.$$

Then expressions (1.94), (1.95) are represented as

$$G_T(x, y) = \exp\left(-\lambda\left(\frac{1}{(x - y)^3} + \frac{1}{2(x + 1)^2} + \frac{1}{2(y + 1)^2} + \right.\right.$$
$$+ \left(\frac{1}{2(x + 1)^2} - \frac{1}{2(y + 1)^2}\right) + \left(\frac{1}{x + 1} - \frac{1}{y + 1}\right) \times$$
$$\times \frac{3}{(x - y)^4} - \frac{1}{2(y + T + 1)^2} - \frac{1}{2(x + T + 1)^2} - \frac{1}{(x - y)^3} \times$$
$$\times \left(\frac{1}{2(x + T + 1)^2} - \frac{1}{2(y + T + 1)^2}\right) - \left(\frac{1}{x + T + 1} + \frac{1}{y + T + 1}\right) \times$$

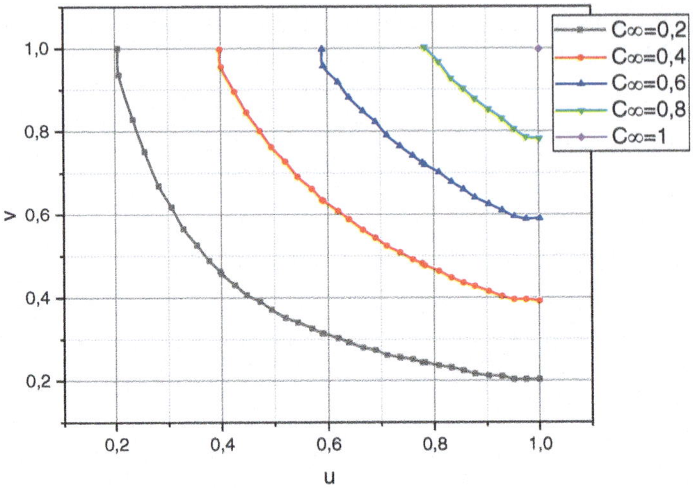

Fig. 1.10 2D representation of the dependence $C_\infty = f(u, v; \lambda = 2, \alpha = 2)$ calculated by expression (1.109)

$$\times \frac{3}{(x-y)^4} + \ln\frac{(y+1)(x+T+1)}{(x+1)(y+T+1)}\Bigg) \times \frac{6}{(x-y)^5}\Bigg),$$
$$x, y \geq 0, 0 < T \leq \infty, x \neq y; \tag{1.111}$$

$$G_\infty(x, y) = \exp\Bigg(-\lambda\Bigg(\frac{1}{(x-y)^3}\Bigg(\frac{1}{2(x+1)^2} + \frac{1}{2(y+1)^2}\Bigg) +$$
$$+ \frac{1}{2(x+1)^2} + \frac{1}{2(y+1)^2} + \Bigg(\frac{1}{x+1} + \frac{1}{y+1}\Bigg) \times$$
$$\times \frac{3}{(x-y)^4} - \ln\Bigg(\frac{x+1}{y+1}\Bigg) \times \frac{6}{(x-y)^5}\Bigg)\Bigg),$$
$$x, y \geq 0, x \neq y.$$

The partial functions of the distribution of the values of the maximum residual time of service in the first and second subsystems $G_{1,T}(x)$, $G_{2,T}(y)$ based on expression (1.111) are represented as

$$G_{1,T}(x) = \exp\Bigg(-\lambda\Bigg(\frac{1}{2(x+1)^2} - \frac{1}{2(x+T+1)^2}\Bigg)\Bigg), \quad x \geq 0,$$
$$G_{2,T}(y) = \exp\Bigg(-\lambda\Bigg(\frac{1}{2(y+1)^2} - \frac{1}{2(y+T+1)^2}\Bigg)\Bigg), \quad y \geq 0. \tag{1.112}$$

Expressions (1.111) and (1.112) at $x = y$ taking the form:

$$G_T(x, y = x) = \exp\left(-\lambda\left(\frac{1}{5(x+T+1)^5} - \right.\right.$$
$$\left.\left. - \frac{1}{(x+T+1)^2} - \frac{1}{5(x+1)^5} + \frac{1}{(x+1)^2}\right)\right), \tag{1.113}$$
$$x \geq 0, 0 < T \leq \infty,$$
$$G_\infty(x, y = x) = \exp\left(-\lambda\left(\frac{1}{(x+1)^2} - \frac{1}{5(x+1)^5}\right)\right), x \geq 0.$$

Taking into account expressions (1.111) and (1.113), expression (1.98) for the definition of the copula $C_\infty(u, v)$ is represented as

$$C_\infty(u, v = u) = u^2 \exp\left(\frac{4\ln^2 u}{5\lambda}\sqrt{\frac{2\ln u}{\lambda}}\right). \tag{1.114}$$

The correlation coefficient (1.99) for the copula (1.114) is defined as

$$\beta_C = 4C_\infty\left(\frac{1}{2}, \frac{1}{2}\right) - 1 = \exp\left(\left(\frac{\lambda}{2\ln 2}\right)^{-\frac{3}{2}} \times \frac{2\ln 2}{5}\right) - 1$$
$$= 2^\wedge\left(\frac{2}{5}\left(\frac{\lambda}{2\ln 2}\right)^{-\frac{3}{2}}\right) - 1. \tag{1.115}$$

The obtained analytical results on the study of situations when the stochastic values X and Y, which characterize the time of service of subrequests in the first and second subsystems, respectively, distributed according to the Pareto distribution law with parameters $\alpha = 2$ and $\alpha = 3$, allow us to describe a generalized situation when parameter $\alpha > 1$:

$$B_1(x) = 1 - (x+1)^{-\alpha}, B_2(y) = 1 - (y+1)^{-\alpha},\ x = y \geq 0,\ \alpha > 1.$$

Then we express expression (1.95) as

$$G_\infty(x, y = x) = \exp\left(-\lambda\left(\frac{2}{(\alpha-1)(x+1)^{\alpha-1}} - \frac{1}{(2\alpha-1)(x+1)^{2\alpha-1}}\right)\right),$$
$$x = y \geq 0. \tag{1.116}$$

Partial functions of distribution of values of the maximum residual service time in the first and second subsystems $G_{1,T}(x) = G_{21,T}(y)\forall x = y$ based on expression (1.116) we will present as

$$G_{1,\infty}(x) = \exp\left(-\lambda \frac{1}{(\alpha - 1)(x + 1)^{\alpha - 1}}\right), \quad x = y \geq 0. \tag{1.117}$$

The function inverse to (1.117) is written as an expression

$$\frac{1}{G_{1,\infty}(u)} = \left(\lambda \frac{1}{(\alpha - 1)\ln u}\right)^{\frac{1}{\alpha-1}} - 1. \tag{1.118}$$

Taking into account expressions (1.115) and (1.118), expression (1.98) for the definition of the copula $C_\infty(u, v)$ is represented as

$$\begin{aligned} &C_\infty(u, v = u) \\ &= u^2 \exp\left(\lambda\left(-\frac{1}{\lambda}(\alpha - 1)\ln u\right)^\wedge \left(\frac{2\alpha - 1}{\alpha - 1}\right) \times \frac{1}{2\alpha - 1}\right) \\ &= u^2 \times u^\wedge\left(\left(-\frac{1}{\lambda}(\alpha - 1)\ln u\right)^\wedge \left(\frac{1}{\alpha - 1}\right) \times \frac{(\alpha - 1)^2 \ln u}{\lambda(2\alpha - 1)}\right). \end{aligned} \tag{1.119}$$

The correlation coefficient (1.99) for the copula (1.119) is defined as

$$\begin{aligned} \beta_C &= 4C_\infty\left(\frac{1}{2}, \frac{1}{2}\right) - 1 \\ &= -1 + \exp\left(\left(\lambda \frac{1}{(\alpha - 1)\ln 2}\right) \wedge \left(-\frac{\alpha}{\alpha - 1}\right) \times \frac{(\alpha - 1)\ln 2}{2\alpha - 1}\right) \\ &= -1 + 2 \wedge \left(\left(\lambda \frac{1}{(\alpha - 1)\ln 2}\right) \wedge \left(-\frac{\alpha}{\alpha - 1}\right) \times \frac{\alpha - 1}{2\alpha - 1}\right). \end{aligned}$$

Assume that the stochastic values of X and Y, which characterize the time of service of subrequests in the first and second subsystems, respectively, are distributed according to the uniform distribution law on the interval [0, 1]:

$$B_1(x) = x, \quad B_2(y) = y, \quad x, y \in [0, 1].$$

Then we express expression (1.95) as

$$\begin{aligned} G_\infty(x, y) = \exp\Bigg(-\lambda\Bigg(\frac{1}{2}(1 - x)^2 + \frac{1}{2}(1 - y)^2 - \frac{1}{6}(1- \\ -(3|x - y| + 2(1 - \max(x, y)))\max(x, y))^2\Bigg)\Bigg), \quad x, y \in [0, 1]. \end{aligned} \tag{1.120}$$

For $T > 1$ and $x \leq y$ we can write

$$\int_0^T (1 - B_1(t+x)B_2(t+y))dt = \int_0^{1-x} (1-(t+x))dt + \int_0^{1-y} (1-(t+x)(t+y))dt$$
$$= \frac{1}{2}(1-x)^2 + \frac{1}{2}(1-y)^2(1+x-y)$$
$$+ -\frac{1}{3}(1-y)^3,$$

and for $T > 1$ and $x > y$ we can write

$$\int_0^T (1 - B_1(t+x)B_2(t+y))dt = \int_0^{1-y} (1-(t+y))dt + + \int_0^{1-x} (1-(t+x)(t+y))dt$$
$$= \frac{1}{2}(1+y-x)(1-x)^2 + \frac{1}{2}(1-y)^2 - \frac{1}{3}(1-x)^3.$$

The partial functions of the distribution of the values of the maximum residual time of service in the first and second subsystems $G_{1,T}(x)$, $G_{2,T}(y)$ based on expression (1.120) are represented as

$$G_{1,T}(x) = \exp\left(-\frac{1}{2}\lambda(1-x)^2\right), G_{2,T}(y) = \exp\left(-\frac{1}{2}\lambda(1-y)^2\right), x, y \in [0, 1]. \tag{1.121}$$

Write the inverse functions to the described expressions (1.121):

$$\frac{1}{G_1(u)} = 1 - \sqrt{-\frac{2}{\lambda}\ln u}, \frac{1}{G_2(v)} = 1 - \sqrt{-\frac{2}{\lambda}\ln v}. \tag{1.122}$$

Taking into account expressions (1.120) and (1.122), expression (1.98) for the definition of the copula is represented as

$$C_\infty(u,v) = uv \times ((u,v)\max) \wedge \frac{1}{\sqrt{\lambda}}$$
$$\times \left(\frac{1}{3}\sqrt{-2\ln(\max(u,v))} - \sqrt{-2\ln(\min(u.v))}\right), \tag{1.123}$$

$$C_\infty(u, v = u) = u^2 \times u^\wedge\left(-\frac{2}{3}\sqrt{-\frac{2}{\lambda}\ln u}\right).$$

The correlation coefficient (1.99) for the copula (1.123) is defined as

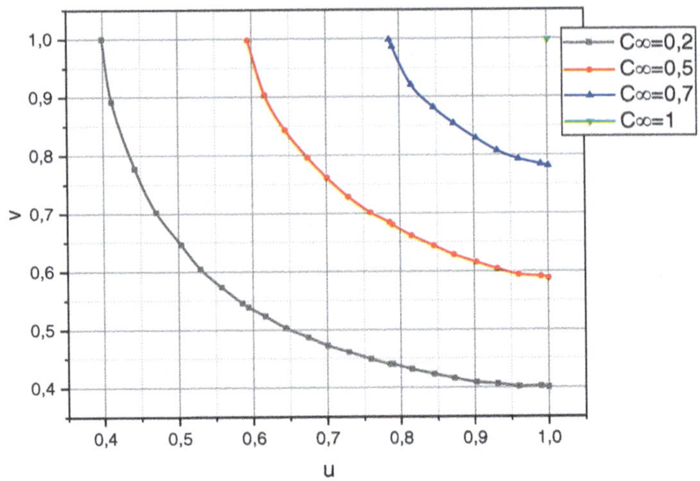

Fig. 1.11 2D representation of the dependence $C_{\infty} = f(u, v; \lambda = 2, [0, 1])$ calculated by expression (1.123)

$$\beta_C = 4C_{\infty}\left(\frac{1}{2}, \frac{1}{2}\right) - 1 = -1 + 2^{\wedge}\left(\frac{2}{3}\sqrt{\frac{2}{\lambda}}\ln 2\right). \tag{1.124}$$

The graphical interpretation of expression (1.123) in the parametric space (u, v) and at the values of the established parameters $\lambda = 2, [0, 1]$ is presented in Fig. 1.11.

Finally, suppose that the stochastic quantities X and Y, which characterize the time of service of subrequest in the first and second subsystems, respectively, are distributed according to the generalized distribution law of the form

$B_1(x) = (1 - x)^{\alpha}, \quad B_1(x) = (1 - y)^{\alpha}, \quad \alpha > 0$,on the interval $[0, 1]$. Let $x = y$. Let us represent expression (1.95) as

$$G_{\infty}(x, x) = \exp\left(-\lambda\left(\frac{2(1 - x)^{\alpha+1}}{\alpha + 1} - \frac{(1 - x)^{2\alpha+1}}{2\alpha + 1}\right)\right).$$

Then we express the partial function as

$$G_{1,\infty}(x) = \exp\left(-\lambda\frac{(1 - x)^{\alpha+1}}{\alpha + 1}\right).$$

Based on typical transformations taking into account these expressions we will formulate expressions for the calculation of a copula:

$$C_\infty(u,u) = u^2 \exp\left(\lambda\left(-\frac{1}{\lambda}(\alpha+1)\ln u\right) \wedge \left(\frac{2\alpha+1}{\alpha+1}\right) \times \frac{1}{2\alpha+1}\right)$$
$$= u^2 \times u \wedge \left(\left(-\lambda\frac{1}{(\alpha+1)\ln u}\right) \wedge \left(\frac{1}{\alpha+1}\right) \times \frac{(\alpha+1)^2 \ln u}{\lambda(2\alpha+1)}\right),$$

and the corresponding correlation coefficient:

$$\beta_C = 4C_\infty\left(\frac{1}{2},\frac{1}{2}\right) - 1$$
$$= -1 + \exp\left(\left(\left(\lambda\frac{1}{(\alpha+1)\ln 2}\right) \wedge \left(\frac{\alpha}{\alpha+1}\right) \times \frac{(\alpha+1)\ln 2}{2\alpha+1}\right)\right)$$
$$= -1 + 2 \wedge \left(\left(\left(\lambda\frac{1}{(\alpha+1)\ln 2}\right) \wedge \left(\frac{\alpha}{\alpha+1}\right) \times \frac{\alpha+1}{2\alpha+1}\right)\right).$$

In Fig. 1.9–3.3 visualized functional dependences $C_\infty = f(u, v)$ are calculated by expressions (1.104), (1.109), (1.123). Now in Fig. 1.12, we present the functional dependencies $\beta_C = f(\rho)$ for exponential, Pareto $\alpha = 2$, Pareto $\alpha = 3$, and uniform distributions, calculated based on expressions (1.105), (1.110), (1.115), and (1.124), respectively. Recall that ρ is the load factor of the studied system, which, when introduced into expressions (1.105), (1.110), (1.115), (1.124), modifies them as follows: $\beta_C = -1 + 2^\wedge(\ln 2/2\rho)$, $\beta_C = -1 + 2^\wedge(\ln^2 2/3\rho^2)$, $\beta_C = -1 + 2^\wedge\left(\ln^2 2\sqrt{\ln 2/\rho}\,/\,3\rho^2\right)$, $\beta_C = -1 + 2^\wedge\left(2\sqrt{\ln 2/\rho}\,/\,3\right)$.

Let us summarize the obtained analytical results of the study of the distribution laws of stochastic quantities X and Y, which characterize the time of service of subrequests in the first and second subsystems, respectively. Suppose that the functions

Fig. 1.12 Dependencies $\beta_C = f(\rho)$ for the corresponding distribution laws $G_T(x.y)$

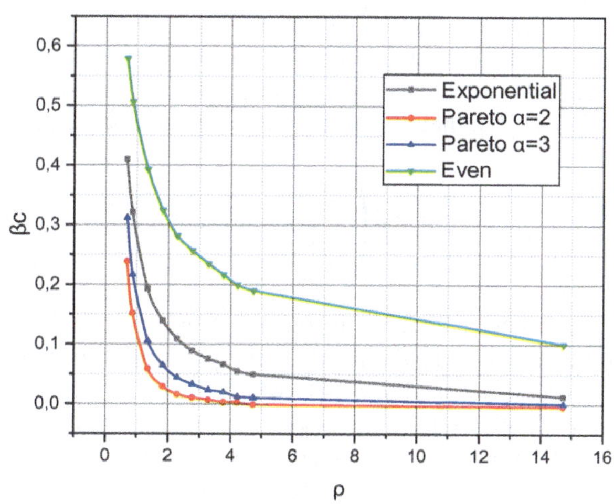

$F_{i,T}, i = \overline{1,2}$ defined by expressions (1.97) belong to the attraction domain of maximally stable distributions H_i. This means that there are such normalizing constants $a_i(s), b_i(s) > 0$ that $F^s_{1,T}(a_1(s)x + b_1(s)) \to H_1(x), F^s_{2,T}(a_2(s)y + b_2(s)) \to H_2(y)$, $s \to \infty$. Accordingly, we can write: $G_T(a_1(\lambda)x + b_1(\lambda), a_2(\lambda)y + b_2(\lambda)) \to H_1^{\mu_{1,T}}(x)H_2^{\mu_{2,T}}(y), \lambda \to \infty$.

From the obtained expressions it follows that at a high load of the studied system the stochastic values that characterize the time of service of subrequests in the subsystems are asymptotically independent. Based on this fact, with the infinite mathematical expectation of the time of service of subrequests in the subsystems: $T \to \infty$, we formulate the condition: if $\overline{B}_1(x) \sim c_1x^{-\alpha_1}$, $\overline{B}_2(y) \sim c_2y^{-\alpha_2}$, $x, y > 0$, $x, y \to \infty$, $c_1, c_2 > 0$, $\alpha_1, \alpha_2 \in (0, 1]$, then for $\alpha_1 = \alpha_2 = 1$: $G_T(Tx, Ty) \to \left(\frac{x}{x+1}\right)^{c_1\lambda}\left(\frac{y}{y+1}\right)^{c_2\lambda}$, and for $0 < \alpha_1, \alpha_2 < 1$: $G_T\left(T^{\frac{1}{\alpha_1}}x, T^{\frac{1}{\alpha_2}}y\right) \to$ $\exp(-c_1\lambda x\verb|^|(-\alpha_1)) \exp(-c_2\lambda y\verb|^|(-\alpha_2))$. Accordingly, the normalized maximum stochastic values of the residual time of service of subrequests in the subsystems are asymptotically independent.

We now investigate the influence of the characteristics of the flow of incoming requests on the maximum stochastic values of the residual time of service of subrequests in the subsystems.

Let the flow of input queries directed to the system under study be nonstationary Poisson with the intensity considered by the function $\lambda = \lambda(t) \geq 0$, $0 < T < \infty$. Then the distribution function $G_T(x.y)$ of the two-dimensional stochastic value of the maximum of values of the residual time of service of subrequests in the subsystems of such a system at a time T is defined as

$$G_T(x, y) = \exp\left(-\int_0^T \lambda(t)(1 - B_1(T - t + x)B_2(T - t + y))dt\right). \quad (1.125)$$

If the flow of input requests directed to the system under study is twice stochastic Poisson with the intensity considered by the stochastic process $\lambda = \lambda(t) \geq 0$, $0 < T < \infty$, then the distribution function $G_T(x.y)$ is defined as

$$G_T(x, y) = \mathrm{E}\exp\left(-\int_0^T \lambda(t)(1 - B_1(T - t + x)B_2(T - t + y))dt\right). \quad (1.126)$$

Expression (1.126) is formed from expression (1.125) if we average all the trajectories of development of the process $\lambda(t)$. Assume that $\lambda(t) = \max(\gamma(t), 0)$, $\gamma(t) = \lambda_0 + \xi(t)$, $\lambda_0, \sigma > 0$, where $\xi(t)$ is the censored at $t \geq 0$ stationary Gaussian process with zero mathematical expectation and the covariance function $R(0) = 1$. Substituting the so-defined stochastic process $\lambda(t)$ into expression (1.126) we obtain

$$G_T(x, y) \leq \exp\left(-\lambda_0 \int\limits_0^T (1 - B_1(t + x) B_2(t + y))dt + \frac{\sigma^2}{2} \int\limits_0^T \int\limits_0^T (1 - B_1(u + x)) \times \right.$$
$$\left. \times R(u - v)(1 - B_2(u + y))(1 - B_2(v + y))(1 - B_1(v + x))dudv\right).$$

Note that the obtained function, which characterizes the intensity of the Gaussian process, is approximate because under such conditions the intensity can potentially be negative. However, the probability of such an event at $\sigma << \lambda_0$ is so small that it can be neglected.

Here are examples of the use of expressions (1.125), (1.126).

Let the time of service of subrequests in subsystems be a stochastic quantity, which is characterized by an exponential distribution (1.100). Then, according to expression (1.117), we define the function $G_T(x \cdot y)$ as

$$G_T(x \cdot y) = \exp(-A_1(T) \exp(-\beta_1 x) - A_2(T) \exp(-\beta_2 y)$$
$$-A_3(T) \exp(-\beta x_1 - \beta_2 y)), \ x, y \geq 0,$$

where

$$A_1(T) = \int\limits_0^T \lambda(T - t) \exp(-\beta_1 t)dt, A_2(T) = \int\limits_0^T \lambda(T - t) \exp(-\beta_2 t)dt,$$

$$A_1(T) = \int\limits_0^T \lambda(T - t) \exp(-\beta_1 - \beta_2)dt.$$

Taking into account expression (1.118), for the exponential distribution, we write

$$\begin{aligned}
&G_T(x, y) \leq \exp(-A_1(T) \exp(-\beta_1 x) - \\
&-A_2(T) \exp(-\beta_2 y) + A_3(T) \exp(-\beta_1 x - \beta_2 y)) + \\
&+ D_1(T) \exp(-2\beta_1 x) + D_2(T) \exp(-2\beta_2 y) + \\
&+ D_3(T) \exp(-\beta_1 x - \beta_2 y) - D_4(T) \exp(-2\beta_1 x - \beta_2 y) - \\
&- D_5(T) \exp(-\beta_1 x - 2\beta_2 y) + D_6(T) \exp(-2\beta_1 x - 2\beta_2 y),
\end{aligned}$$

where $x, y \geq 0$, $A_1(T) = \frac{\lambda_0}{\beta_1}(1 - \exp(-\beta_1 T))$, $A_2(T) = \frac{\lambda_0}{\beta_2}(1 - \exp(-\beta_2 T))$, $A_3(T) = \frac{\lambda_0}{\beta_1 + \beta_2}(1 - \exp(-\beta_1 T - \beta_2 T))$,

$$D_1(T) = \frac{\sigma^2}{2} \int\limits_0^T \int\limits_0^T R(u - v) \exp(-\beta_1 u - \beta_1 v)dudv,$$

$$D_2(T) = \frac{\sigma^2}{2}\int_0^T\int_0^T R(u-v)\exp(-\beta_2 u - \beta_2 v)dudv,$$

$$D_3(T) = \frac{\sigma^2}{2}\int_0^T\int_0^T (\exp(-\beta_1 u - \beta_2 v) + \exp(-\beta_2 u - \beta_1 v))R(u-v)dudv,$$

$$D_4(T) = \frac{\sigma^2}{2}\int_0^T\int_0^T (\exp(-(\beta_1+\beta_2)u - \beta_1 v) + \exp(-\beta_1 u - (\beta_1+\beta_2)v))R(u-v)dudv,$$

$$D_5(T) = \frac{\sigma^2}{2}\int_0^T\int_0^T R(u-v)(\exp(-\beta_2 u - (\beta_1+\beta_2)v) + \exp(-(\beta_1+\beta_2)u - \beta_2 v))dudv,$$

$$D_6(T) = \frac{\sigma^2}{2}\int_0^T\int_0^T R(u-v)(\exp(-(\beta_1+\beta_2)(u+v)))dudv.$$

For example, for $R(t) = \exp(-\delta|t|)$, $\delta > 0$, the functions $D_1(t) \div D_6(t)$ at $t \to \infty$ will take the following analytical form:

$$D_1(\infty) = \frac{\sigma^2}{2\beta_1(\delta+\beta_1)},\ D_2(\infty) = \frac{\sigma^2}{2\beta_2(\delta+\beta_2)},$$

$$D_3(\infty) = \frac{\sigma^2(\beta_1+\beta_2+2\delta)}{(\delta+\beta_1)(\delta+\beta_2)(\beta_1+\beta_2)},$$

$$D_4(\infty) = \frac{\sigma^2(2\beta_1+\beta_2+2\delta)}{(\delta+\beta_1+\beta_2)(\delta+\beta_1)(2\beta_1+\beta_2)},$$

$$D_5(\infty) = \frac{\sigma^2(\beta_1+2\beta_2+2\delta)}{(\delta+\beta_1+\beta_2)(\delta+\beta_2)(\beta_1+2\beta_2)},$$

$$D_6(\infty) = \frac{\sigma^2}{2(\delta+\beta_1+\beta_2)(\beta_1+\beta_2)}.$$

Therefore, the section comprehensively describes such a significant parameter for estimation of the functional safety of DC-like systems of type $M^{(2)}\big|G_2|\infty$ as the maximum residual time of service of subrequests in the subsystems. Analytical dependencies are formalized for a variant of a queuing system consisting of two parallel subsystems with an infinite number of service devices. The functioning of such a system has been studied both in stationary mode and in the conditions of the censored time. The connection between the mentioned characteristic parameter and functional security is obvious because the investigated parameter characterizes the time required to complete the processing of subrequests by all service devices of both subsystems after turning off the flow of incoming requests. The flow of incoming requests was characterised by its intensity, which was assumed to be either time-independent, time-dependent, or governed by a stochastic (stationary Gaussian) process. The stochastic nature of the studied characteristic parameter is taken into account in the context of such distribution laws as exponential, hyperexponential, Pareto, uniform. These studies are generalized in the form of appropriate analytical expressions for the calculation of the copula and the correlation coefficient. Analytically, it was found that at a high level of load of the studied system or at the infinite mathematical expectation of time of service, the stochastic values that characterize the residual time of service of subrequests in the subsystems of such a system are asymptotically independent.

Studies have also revealed the relationship between the distribution law of the characteristic parameter and the rate of decline of the correlation coefficient in terms of increasing the intensity of the flow of incoming requests. The revealed dependence has a degree character: $\beta_C = f\left(1/\lambda^s\right)$. Accordingly, for

- Pareto distribution $\left(B(x) = 1 - 1/(x+1)^\alpha, x \in [0,1], \alpha > 0\right)$: $s = \alpha/(\alpha - 1)$, where $s \to \infty$ at $\alpha \to 1$ and $s \to 1$ at $\alpha \to \infty$;
- exponential distribution: $s = 1$;
- uniform distribution: $s = 1/2$ at $\alpha = 1$ and $s \to 1$ at $\alpha \to \infty$;
- distributions of the form $B(x) = 1 - (x+1)^\alpha, x \in [0,1], \alpha > 0$: $s = \alpha/(\alpha + 1)$, where $s \to 0$ at $\alpha \to 0$.

Therefore, the greater the difference in the values of time of service of subrequests in the subsystems, the smaller the relationship between the maximum residual values of time of service in the subsystems and the synchronicity of incoming requests. The mathematical apparatus presented in the section can increase the efficiency of functioning of the information systems presented in [100–104].

1.5 Conclusion

The presence of society in the Information Age is accompanied by constant challenges, the consequences of inadequate response to which are becoming increasingly severe. Under such conditions, only profile mathematical modelling is a tool that allows us to formulate strategies for the effective use and development of information systems of various scales and purposes. A promising platform for such modelling is the «Divide and Conquer» paradigm. This thesis was confirmed in the form of numerous scientific studies, the results of which are summarized in the first section. Current models of DC-like systems and methods of accurate and approximate calculation of their characteristic parameters, in particular, the time of system response, were analysed. At the same time, the main part of this research is devoted to the presentation of models of DC-like systems in the taxonomy of such attributes of dependability as confidentiality (second and third sections) and functional safety (fourth section).

Section 1.2 examines an instance of DC-like systems that operated under probable cyber threats (requests of the first type). The implementation of the cyber threat was modelled by a mechanism of stochastic update, the activation of which leads to the loss of all ordinary requests (requests of the second type) in the service station of the system. The section formulates expressions for calculating the stationary probability characteristics of the corresponding queuing system, in particular, the probability of downtime, the probability of servicing an ordinary request (in terms of the generating function), the mathematical expectation of the number of ordinary requests in the system.

Section 1.3 proposes a mathematical apparatus for formalizing systems of equilibrium equations, which characterize the process of two-factor authentication as a network consisting of a tuple of two sequentially connected queuing systems with mechanisms of generalized update. The calculation of the distribution of the time spent by the request in such a network is formalized. The latter was modelled by a Markov chain with a finite state space. The process of calculating the mentioned characteristic parameter is defined in terms of the Laplace–Stieltjes transformation, because the value of the time between the transitions of the nested Markov network was described, albeit by different, but exponential distributions. However, the obtained mathematical apparatus is characterized by a relatively high computational complexity, which prompts the representation of the calculation of the time characteristics of such networks in terms of Little's theorem. In such circumstances, relying on the conducted research [105–109], the author plans to target further research on finding more efficient computational methods for solving the corresponding systems of equilibrium equations.

Section 1.4 comprehensively describes such a significant parameter for estimating the functional safety of DC-like systems as the maximum residual time of service of subrequests in subsystems. In particular, this characteristic parameter is analytically formalized for a variant of the queuing system consisting of two parallel operating subsystems with an infinite number of service devices. The connection between this characteristic parameter and functional security is obvious because the first characterizes the time required to complete the processing of subrequests by all service devices of both subsystems after turning off the flow of incoming requests. The functioning of such a system has been studied both in stationary mode and in the conditions of the censored time. The flow of incoming requests was characterised by its intensity, which was assumed to be either time-independent, time-dependent, or governed by a stochastic (stationary Gaussian) process. The stochastic nature of the studied characteristic parameter is taken into account in the context of such distribution laws as exponential, hyperexponential, Pareto, uniform. These studies are generalized in the form of appropriate analytical expressions for the calculation of the copula and the correlation coefficient.

Chapter 2
DC-Like Systems with BMAP-Flow as a Realistic Platform for Modelling Information and Communication Processes

2.1 Introduction

DC-like systems serve as a crucial tool for modelling processes in technical, industrial, economic, administrative, and other complex information systems, where managing request flows under constrained resources is required. The stochastic nature of request arrivals, along with the necessity for efficient allocation of computational, informational, or physical resources, leads to delays and service rejections. The theory of queuing systems [110–114] has been developed to minimize such losses by formalizing service processes, constructing mathematical models, and optimizing the operational parameters of DC-like systems.

Modern approaches to modelling information and communication processes are based on the concept of distributed systems, within which various architectural solutions are applied, including request processing based on principles of partitioning and synchronization, independent processing, or multi-channel servicing, among others. A crucial aspect of analysing such service systems involves examining temporal characteristics, assessing load distribution, and developing balancing mechanisms that enhance the efficiency of modelled objects under varying request flows.

The development of the theoretical foundations for modelling service systems is linked to the work of Danish mathematician and engineer A. Erlang, published between 1910 and 1922 [115]. His research focused on solving problems related to the design of telephone networks and laid the foundation for the mathematical modelling of request servicing processes in DC-like systems with limited resources. These studies introduced a Poisson model for the request flow arriving at a network node. This model characterizes the incoming request flow as a stationary, ordinary, memoryless process or as a recurrent flow with an exponential distribution of inter-arrival intervals. Additionally, it was assumed that the service time of a request also follows an exponential distribution. Given the memoryless property of the exponential distribution, the number of requests in such service systems was described

V. Kovtun, *Formal Models of Divide-and-Conquer Queuing Systems in 5G Infrastructures*, https://doi.org/10.1007/978-3-032-06187-4_2

by a one-dimensional Markov process, which allowed for the analytical determination of its steady-state distribution. Despite the somewhat simplified nature of this assumption, Erlang's theoretical results aligned well with empirical measurements in real telephone networks. The further development of this concept was substantiated by the research of Khinchin and Grigelionis [116–118], who demonstrated that under the condition of uniformly low individual flow intensities, their superposition approximates a stationary Poisson flow. This result became a probabilistic analogue of the central limit theorem, which states that under certain conditions, the sum of a large number of independent random variables converges to a normal distribution. Similarly, request flows arriving at a service system node represent the superposition of numerous independent low-intensity flows generated by individual subscribers. Consequently, the stationary Poisson flow model proved to be an effective tool for describing request servicing dynamics in telephone networks and, later, in a broader class of information and communication systems. Further research into information and communication processes has led to an expansion of the nomenclature of classical DC-like models. In particular, modelling methods have been developed for service systems that account for resource limitations, the stochastic nature of incoming request processing, and the specific features of modern information and communication platform architectures. This has facilitated the emergence of new mathematical approaches, such as queuing networks, parallel and distributed service systems, as well as order statistics methods and matrix-analytic techniques, which enable high-precision load forecasting and optimization of modern communication system performance.

The development of computer data transmission networks based on packet switching, as opposed to traditional circuit switching in analogue telephone channels, necessitated the creation of new mathematical tools for their optimal design. Similar to the approaches proposed by A. Erlang for studying telephone networks, the early research of Kleinrock et al. [119–122] on data transfer processes in computer networks relied on Poisson flow models and the exponential distribution of packet transmission duration. This made it possible to apply existing queuing theory methods to evaluate network performance, design topological structures, manage routing, and select optimal network protocol parameters. As computer networks grew in complexity, the need arose to refine these models to account for the stochastic nature of traffic, the limitations of communication resources, the possibility of service prioritization, and the implementation of adaptive traffic management mechanisms. These pressing needs prompted the development of new approaches, including queuing networks with branched structures, hybrid architectures for multi-channel servicing, adaptive traffic management algorithms, and methods for optimal resource allocation in packet-switched networks. The further advancement of theoretical foundations for computer network modelling was marked by the introduction of matrix-analytic methods, which enabled the resolution of equilibrium equations in complex DC-like systems, and order statistics methods, which provided more precise estimates of delay duration and throughput capacity. Significant contributions to these fields were made by Little et al. [123–127]. The application of these approaches facilitated more effective load forecasting and the optimization of interactions between information system nodes, laying the foundation for the continued development of modern information and communication technologies.

The further development of the theory of information and communication systems was driven by the emergence of Integrated Services Digital Networks (ISDN) [128, 129], which represented an evolutionary extension of early packet-switching technologies. The primary characteristic of these networks was the use of a unified infrastructure of hardware and software tools for processing and transmitting various types of multimedia content, including voice calls, streaming data, large information arrays, text messages, and video content. The introduction of ISDN marked a significant step towards the unification of information and communication technologies and the enhancement of network resource efficiency. However, the high heterogeneity of information flows in ISDN networks introduced new challenges for mathematical modelling, as data streams tend to exhibit instability and uneven load distribution (commonly referred to as bursty traffic). Such traffic patterns are characterized by high temporal variability, leading to significant fluctuations in the load on communication channels. Moreover, ISDN networks exhibit a correlation between request volumes within specific time intervals, which can persist even across distant time frames. The increasing complexity of traffic flows has become a critical factor in the development of next-generation broadband networks. In particular, modelling data transmission processes in 4G networks required accounting for correlated loads, stochastic delays, and adaptive mechanisms for managing categorized flows.

The further evolution of this concept was reflected in theoretical research that preceded the development of modern 5G networks, designed to support an even broader range of services, including massive IoT connectivity, autonomous systems, and mission-critical applications with ultra-low latency. However, the advancement of communication technologies does not stop at 5G. The ongoing deployment of sixth-generation (6G) networks presents researchers with even more complex challenges, including optimizing the use of spectral and computational resources, integrating artificial intelligence and distributed computing technologies, and ensuring ultra-reliable communication with extremely low latency. Key characteristics of 6G networks will include terahertz communication bands, quantum cryptography, the widespread use of intelligent surfaces for signal control, and holographic communications. The need for analytical modelling of processes specific to 6G networks necessitates the development of new mathematical approaches, including queuing networks with variable request intensity, Markov processes with long-term dependencies, and matrix-analytic methods that enable the assessment of load dynamics and the optimization of traffic management mechanisms. In the context of 6G, future research perspectives are focused on developing efficient models for network load forecasting, optimizing energy consumption, and ensuring the effective utilization of dynamic network topologies with a high degree of autonomy.

One of the key characteristics of modern traffic in information and communication networks is its complex, nonlinear dynamics, which cannot be effectively modelled using classical Poisson processes. One approach to describing such behaviour involves the use of self-similar traffic flows [131–133], which account for the presence of long-term dependencies between events and significant deviations from stationarity. These flows exhibit fractal properties and display correlated variability across different time scales, a phenomenon typical of many information and communication systems, including Internet traffic, streaming multimedia transmissions, and networks with limited bandwidth.

Despite their theoretical alignment with real-world processes in information and communication systems, the application of self-similar traffic models in *DC*-like systems presents several challenges. In particular, self-similar traffic is extremely challenging to formalize, making its integration into analytical network performance models more complex. The most effective approach to its representation involves the superposition of on/off sources, where periods of activity and inactivity follow heavy-tailed distributions [133–136]. However, this method relies on asymptotic assumptions, which limit the ability to obtain precise analytical estimates of the operational parameters of the studied service systems. The analysis of DC-like systems operating with self-similar traffic is further complicated by the fact that the incoming flow is only one component of such systems. Within the framework of classical approaches, effective analytical research aims to derive the simplest possible mathematical description that does not lead to excessive computational complexity. Accounting for the self-similarity of incoming request flows, on the contrary, significantly increases computational resource requirements and necessitates the use of numerical modelling or simulation-based approaches, such as traffic modelling based on fractional Brownian motion, Lévy process methods, or hybrid stochastic models with Markovian modifications [137–142]. Prospective research in this field focuses on developing simplified yet sufficiently accurate approaches for the analytical representation of self-similar traffic. Among these methods, fractal Markov models [143, 144] stand out, as they preserve the key properties of self-similarity while enabling efficient mathematical analysis of service parameters. Further advancements in this domain could substantially improve load forecasting in sixth-generation broadband networks, optimize communication resource management, and enhance the adaptability of information and communication systems to unstable incoming traffic flows. Thus, one of the simplest models of incoming traffic in DC-like systems is the stationary Poisson flow, which is characterized solely by the request arrival intensity. The value of this parameter corresponds to the expected number of events per unit time and simultaneously serves as the parameter of the exponential distribution governing inter-arrival periods. The simplicity of this model ensures its widespread use in classical analytical studies but also imposes several limitations. In particular, a stationary Poisson flow exhibits a fixed coefficient of variation, an inter-event interval duration of unity, and zero correlation between consecutive arrival intervals. This means that if statistical analysis of empirical data in real information and communication systems reveals significant deviations of these parameters from the mentioned values, the Poisson model cannot adequately describe the behaviour of the incoming traffic flow. Thus, the need arose for the development of new mathematical models capable of accounting for the stochastic variability and correlation structure of requests in the incoming flow while preserving the ability to obtain analytical results comparable to classical models. The expansion of input flow models began with the consideration of variable request arrival intensity. One of the initial steps in this direction was the Interrupted Poisson Process (IPP) [145, 146], where periods of active request arrivals alternate with time intervals during which no requests are received. The further development of this concept led to the creation of the Switched Poisson Process (SPP) [147, 148], in which the flow intensity switches between two levels according to a predefined process. The next stage in generalizing these models involved the introduction of the Markov Modulated Poisson Process (MMPP) [149, 150], where the flow intensity varies depending on the state of a Markov process with a finite number of states. This model proved to be highly flexible for describing information and communication systems with adaptive traffic. Its further generalization resulted in the formalization of the Batch Markovian Arrival Process (BMAP) [151–154], which allows for the modelling of simultaneous request arrivals in batches at each state transition of the system under study.

The analysis of the adequacy of DC-like models for modern information and communication systems indicates that neglecting correlation dependencies in incoming request flows can lead to significant errors in performance forecasting. At the same time, studies of real information and communication networks and simulations of their behaviour confirm that the presence of positive correlation between consecutive requests significantly deteriorates operational performance. Specifically, it leads to an increase in average waiting time, a higher probability of service denial, and greater strain on network resources. Notably, even with identical average request arrival rates, identical service time distributions, and the same buffer capacity, the probability of request loss in correlated flows can be several orders of magnitude higher compared to uncorrelated flows. Thus, if empirical data indicate the presence of correlation dependencies in the incoming request flow of a target information and communication system, modelling it using a recurrent or, even more so, a stationary Poisson flow is unacceptable. In such cases, the application of models with correlated flows is empirically justified. Significant contributions to the development of the theory of DC-like systems with correlated flows have been made by M. Neuts, D. Lucantoni, V. Ramaswami, S. Chakravarthy, F. Buzzetta, J. Gross, F. Kelly, T. Bonald, M. Haas, R. Combes, P. Johansson, and others. Their research [155–157] has been widely applied in the analysis of modern high-speed information and communication systems, including.

- Broadband wireless networks operating under IEEE 802.11, 802.16, and 802.15 standards in distributed access mode.
- Hybrid high-speed communication systems that integrate laser and radio frequency technologies to ensure reliable data transmission in dynamic environments.
- Cellular mobile communication networks based on 3G, UMTS, LTE (long-term evolution), and 5G standards, where correlated incoming flows play a crucial role in congestion prediction and bandwidth allocation.
- Sensor networks within the Internet of Things, characterized by clustered and highly correlated requests from service stations exchanging data in real time.

- Next-generation optical and satellite networks, where traffic characteristics fluctuate due to weather conditions, atmospheric dynamics, and variations in data sources.

Studies on the operational performance of 5G networks confirm [158–162] that traditional traffic modelling approaches cannot fully account for the stochastic phenomena in data flows generated by technologies such as Ultra-Dense Networks (UDN), Massive IoT, and Ultra-Reliable Low Latency Communications (URLLC). Correlated incoming flows in such networks arise due to nonlinear dependencies in the distribution of user requests, which is particularly relevant in the context of adaptive communication resource allocation and on-demand multiplexing (network slicing). The transition to 6G networks presents even greater challenges in managing correlated flows. It is expected that terahertz communication channels, next-generation satellite networks, and cognitive radio networks will require new approaches to processing correlated requests.

The historical development of the theory of single-line DC-like systems with *BMAP*-flow was thoroughly examined in the works of Lucantoni [163–165]. Further progress in this field by the end of the twentieth century was summarized in a foundational review by Chakravarthy [166], which covered a broad range of models and their applications. Since then, research has expanded significantly, encompassing complex multi-channel sensor systems, network architectures with correlated incoming flows, and their use in modelling load distribution in modern information and communication systems. This section aims to systematize and generalize the latest research findings on DC-like systems with flow, as well as to analyse their applications in information and communications, cloud computing, and adaptive distributed systems.

The content of this section is structured into five subsections:

- Section 2.2 presents the definition of *BMAP*-flow, describes its key properties, and explores its relationship with classical input flow models, including stationary Poisson flows and Markovian Arrival Processes (MAP).
- Section 2.3 provides a general overview of two-dimensional Markov chains used for analysing Markovian service systems. In particular, it examines matrix-analytic methods, which enable an efficient study of the characteristics of complex DC-like systems.
- Section 2.4 explores single-line service system models with an arbitrary service time distribution. Various architectures are considered, including systems with unlimited or finite buffers and systems with retrial mechanisms.
- Section 2.5 presents research on multi-channel DC-like systems with phase-distributed incoming request flows. The analysis covers models with finite and infinite buffers, as well as retrial mechanisms in cases of system overload.
- Section 2.6 examines custom network service systems with correlated request flows, highlighting tandem models, unreliable DC-like systems, and controlled service systems, along with their application in performance analysis of broadband mobile networks, particularly 5G.

2.2 BMAP-Flow as a Derivative of Classical Flow Models

An arbitrary *BMAP*-flow of requests in continuous time is governed by an irreducible Markov chain $\rho_t, t \geq 0$, with a finite state space $\{0, \ldots, S\}$. In the Markov chain ρ_t, it is possible to transition from any state to any other state within a finite number of steps with nonzero probability. The duration of time spent in any state ρ of chain ρ_t follows an exponential distribution with parameter μ_ρ, $\rho \in \overline{0, S}$. Upon completion of the duration of the chain ρ_t in the current state, the *BMAP*-flow with probability $q_n(\rho, \rho')$ transitions to another state $\rho' \in \{\overline{0, S}\}$, $\rho' \neq \rho$. This transition is accompanied by the generation of a batch of n requests, $n \geq 0$. The process can be formalized analytically as follows:

$$q_0(\rho, \rho) = 0,$$

$$\sum_{\rho'=0, \rho' \neq \rho}^{S} q_0(\rho, \rho') + \sum_{n=1}^{\infty} \sum_{\rho'=0}^{S} q_n(\rho, \rho') = 1, \rho \in \overline{0, S}.$$

The parameters characterizing the *BMAP*-flow are typically represented as square matrices A_n, $n \geq 0$, of order $\overline{S} = S + 1$. The elements of matrices A_n are defined as follows:

$$(A_0)_{\rho,\rho'} = \begin{cases} -\mu_\rho \forall \rho = \rho', \\ \mu_\rho q_0(\rho, \rho') \forall \rho \neq \rho', \end{cases}$$

$$(A_n)_{\rho,\rho'} = \mu_\rho q_n(\rho, \rho'), n > 0.$$

As a result, the operation $n > 0$ produces a matrix whose (ρ, ρ')-th element represents the probability that, within the interval $(t, t + dt)$, a batch of n incoming requests will arrive, and at the moment t, the current state ρ of the Markov chain ρ_t, $t \geq 0$, which characterizes the studied *BMAP*-flow, will transition to state ρ' at time $t + dt$. Without loss of generality, we assume $t = 0$.

Instead of storing the set of matrices A_n, $n \geq 0$, it is more efficient to operate with their generating function of the form $A(s) = \sum_{n=0}^{\infty} A_n s^n$.

The matrix $A(1)$ serves as the generation operator for the ρ_t process, $t \geq 0$. The row vector π of the stationary distribution of the Markov chain ρ_t, $t \geq 0$, is the unique solution to the system of linear equations of the form $\pi A(1) = \vec{0}$, $\pi \vec{1} = 1$, where $\vec{0}$ is a row vector of the corresponding length, with all elements equal to 0, and $\vec{1}$ is a column vector of the corresponding length, with all elements equal to 1.

In the context of the introduced notations, the average arrival rate of incoming request batches μ_b is defined as $\mu_b = \pi(A(1) - A_0)\vec{1} = -\pi A_0 \vec{1}$. Accordingly, the average arrival rate of incoming requests in the *BMAP*-flow μ is defined as $\mu = \pi A'(1)\vec{1}$, where $A'(1)$ is the derivative of the matrix generating function $A(s)$ at point $s = 1$.

An arbitrary stationary stochastic flow can, in the first approximation, be characterized by the numerical flow function $N(k, t)$, which represents the probability that k incoming requests will arrive for servicing in a DC-like system within the time interval t. For a stationary Poisson flow, the function $N(k, t)$ is defined as $N(k, t) = (\mu t)^k \exp(-\mu t)/k!$.

At the same time, defining a numerical function $N(k, t)$ for the *BMAP*-flow in an analogous manner is not feasible, as the number of incoming requests arriving during an interval of duration t depends on the state of the Markov chain ρ_τ, $\tau \geq 0$, at the start of the interval t. However, it is possible to compute the probability $Q_{\rho,\rho'}(k, t)$ that within an interval of duration t, k incoming requests will arrive, considering $\rho_t = \rho'$ if $\rho_0 = \rho$. The composition of matrices $Q_{\rho,\rho'}(k, t)$ serves as the foundation for defining the matrix $Q(k, t)$ of order $\overline{S}$, $k \geq 0$. For matrices $Q(k, t)$, we define a system of differential equations of the form $Q'(k, t) = \sum_{n=0}^{k} Q(n, t)A_{k-n}$, $k \geq 0$. The generating function $Q(s, t)$ for matrices $Q(k, t)$ is given by

$$Q(s, t) = \sum_{k=0}^{\infty} Q(k, t)s^k = \exp(A(s)t).$$

In this context, the probability of no incoming requests arriving during an interval of duration t is defined as $Q(0, t) = \exp(A_0 t)$.

The desired matrices $Q(k, t)$, $k > 0$, can potentially be expressed through the matrix generating function $Q(s, t)$ as follows:

$$Q(k, t) = \frac{1}{k!} \left. \frac{\partial^k Q(s, t)}{\partial s^k} \right|_{s=0}, \quad k \geq 0.$$

However, this formula has not found practical application. Instead, the exponential approximation procedure of the Markov process [167] is used to determine the matrices $Q(k, t)$, $k > 0$. Let us introduce the generator G of the transition probabilities of the Markov chain in continuous time. With the use of the generator G the transformation

$$\begin{aligned} \exp(Gt) &= \exp(gt(K - E)) = \exp(-gt)\exp(htK) = \\ &= \sum_{i=0}^{\infty} \exp(-gt)(gtK)^i / i! \end{aligned}$$

becomes attainable, where $g = \max_{j=\overline{0,S}}(-G)_{jj}$, $K = E + G/g$, and E is the identity matrix.

The advantage of this procedure lies in its convergence to the product of a scalar function of the argument t and a matrix K, which is independent of t. A similar decomposition can be applied to the matrix exponent G, provided that its off-diagonal elements are positive, its diagonal elements are negative, and the row sums are either negative or equal to zero. These conditions hold, in particular, for the matrix $\exp(A_0 t)$.

If the parameter $\gamma = \max_{j=\overline{0,S}}(-A_0)_{jj}$ is introduced, the equation $Q(0,t) = \exp(A_0 t)$ can be represented as $Q(0,t) = \sum_{i=0}^{\infty} \exp(-\gamma t)(\gamma t)^i L_0^{(i)} \big/ i!$, where $L_0^{(i)} = \left(E + A_0 \big/ \gamma\right)^i$; the matrices $Q(k,t)$, $k > 0$, can be expressed as $Q(k,t) = \sum_{i=0}^{\infty} \exp(-\gamma t)^i (\gamma t)^i L_k^{(i)} \big/ i!$, where $k > 0$, $i \geq 0$; and the matrices $L_k^{(i)}$ satisfy the following system of recursive relations:

$$L_0^{(0)} = E, L_k^{(0)} = 0, L_0^{(i+1)} = L_0^{(i)} \left(E + A_0 \big/ \gamma\right),$$

$$L_k^{(i+1)} = \frac{1}{\gamma} \sum_{j=0}^{k-1} L_j^{(i)} A_{k-j} + L_k^{(i)} \left(E + A_0 \big/ \gamma\right), k > 0, i \geq 0.$$

When analysing a DC-like system with a *BMAP*-flow, the following parameters are particularly informative:

- Matrix $\exp(A_0 t) A_n dt$, whose $\left(\rho, \rho'\right)$-th element represents the probability that at time $t = 0$, the Markov chain ρ_t, $t \geq 0$, which governs the *BMAP*-flow, was in state ρ, and during the interval $(t, t + dt)$, a batch of n incoming requests arrived, as a result of which the controlling chain changed its state to ρ' at time $t + dt$.
- Matrix $\int_0^{\infty} \exp(A_0 t) A_n dt = -A_n \big/ A_0, n > 0$, whose $\left(\rho, \rho'\right)$-th element represents the probability that at time $t = 0$, the Markov chain ρ_t, $t \geq 0$, which governs the *BMAP*-flow, was in state ρ, and subsequently, a batch of n incoming requests arrived, resulting in the controlling chain transitioning to state ρ'.
- Vector s $\pi A_n \big/ \mu_b$, $n > 0$, which characterizes the probability distribution of the states of the Markov chain ρ_t, $t \geq 0$, governing the *BMAP*-flow, at any arbitrary moment after the arrival of a batch of n incoming requests.
- Probability $Q_n = \pi n A_n \vec{1} \big/ \pi \sum_{m=1}^{\infty} m A_m \vec{1} = \frac{\pi A_n \vec{1}}{\pi A'(1)\vec{1}} n = \frac{\pi A_n \vec{1}}{\mu} n$, $n > 0$, representing the likelihood that a given incoming request will arrive as part of a batch of size n.
- Probability $P_n = \pi A_n \vec{1} \big/ \pi \sum_{m=1}^{\infty} A_m \vec{1} = \frac{\pi A_n \vec{1}}{\pi A_0 \vec{1}} n = \frac{\pi A_n \vec{1}}{\mu_b} n$, $n > 0$, representing the likelihood that a given batch of incoming requests in the *BMAP*-flow has a size of n.
- Vector $\pi \sum_{n=1}^{\infty} A_n \big/ \mu_b = \pi (A(1) - A_0) \big/ \mu_b = -\pi A_o \big/ \mu_b$, which characterizes the probability distribution of the states of the Markov chain ρ_t, $t \geq 0$, governing the *BMAP*-flow, immediately after the arrival of a batch of incoming requests.
- Column vector $\sum_{n=1}^{\infty} n Q(n,t) \vec{1} = A'(1) t \vec{1} + \frac{(\exp(A(1)t) - E - A(1)t)}{A(1) - \pi \vec{1}} A'(1) \vec{1}$, whose ρ-th element represents the average number of incoming requests that arrived during a time interval of duration $(0, t)$. At the same time, at moment $t = 0$, the Markov chain ρ_t, $t \geq 0$, governing the *BMAP*-flow, was in state ρ.
- Average duration $\theta_b = -\pi A_0 \int_0^{\infty} \exp(A_0 t) dt \vec{1} \big/ \mu_b = \frac{1}{\mu_b}$ of the interval between the arrivals of two consecutive batches of requests.

- Moment $\theta_b^{(l)} = \pi(A(1) - A_0)\int_0^\infty \exp(A_0 t)t^l \sum_{n=1}^\infty A_n dt\vec{1} \Big/ \mu_b = l!\pi(-A_0)^{-l+1}\vec{1} \Big/ \mu_b$ of order l of the distribution of the durations of intervals between the arrivals of request batches.
- Variance

$$V = \theta_b^{(2)} - \theta_b^2 = \left(\frac{2\mu_b \pi \vec{1}}{-A_0} - 1\right) \Big/ \mu_b^2 \tag{2.1}$$

of the duration of intervals between the arrivals of request batches.

- Correlation coefficient

$$r = \left(-\frac{\pi A(1)}{\mu_b A_0} - A_0\right)\left(-\frac{\vec{1}}{A_0} - \frac{1}{\mu_b^2}\right) \Big/ V \tag{2.2}$$

the durations of two consecutive intervals between the arrivals of batches of incoming requests.

By adjusting the order and elements of the matrices A_n, $n \geq 0$, it is possible to achieve any predefined values of the coefficient of variation and correlation coefficient. In particular, this enables the reproduction of the characteristics of a real *BMAP*-flow based on the statistical analysis of observations. This approach is particularly useful in cases where traditional models, such as the stationary Poisson flow, fail to adequately capture the empirical patterns of the observed flow. It is important to note that the ability to control these parameters ensures flexibility in stochastic process modelling and allows for the adaptation of the model to the specific conditions of the analysed information and communication system.

In recent years, there has been growing interest in the study of service systems in discrete time. In such models, time is divided into slots of variable duration, and all events (such as the arrival of incoming requests and the completion of their servicing) occur only at the boundaries of these slots. This approach reduces analytical complexity and allows for the modelling of real-world protocols with time constraints. Given this trend, an analogue of the *BMAP*-flow for discrete time has been developed [168, 169], enabling the accurate reproduction of flows with batch arrivals of requests. However, in this section, we focus exclusively on the study of service systems in continuous time, as they offer greater universality.

Alongside the *BMAP*-flow, the Marked Markovian Arrival Process (MMAP) [170–172] is frequently mentioned, as it is used to model complex flows whose elements can be marked according to their type, category, class, or other characteristic parameter. By leveraging the developments related to the *BMAP*-flow, the obtained models can be easily adapted for analysing the *MMAP*-flow. This requires an understanding that, in the context of the *BMAP*-flow, the matrix A_n, $n \geq 0$, represents the transition intensities of the controlling process, which are accompanied by the arrival of request batches of size n. In the context of the *MMAP*-flow, the same matrix represents the transition intensities of the controlling process, which are accompanied by the arrival of requests of a specific n-th type. This adaptation allows for the generalization of *BMAP*-models to analyse heterogeneous flows that include requests of different nature and priority levels. Such models are widely applied in network traffic optimization, distributed computing, and data flow management in information and communication systems.

The successful application of service system analysis results with the *BMAP*-flow for performance evaluation and optimization of real information systems requires the prior approximation of empirical flows in accordance with the *BMAP*-model. This is a crucial stage of research, as real information flows often exhibit complex stochastic structures, including dependencies between requests, temporal variations, and batch arrivals. The process of constructing an *BMAP*-model based on real-world data includes the following stages:

- Identification of key statistical characteristics of the flow, such as average intensity, coefficient of variation, and correlation between request arrival intervals.
- Estimation of the parameters of matrices A_n, $n \geq 0$, enabling the reproduction of empirically observed properties.
- Validation of the adequacy of the resulting *BMAP*-model, including a comparison of empirical and theoretical flow characteristics.

This task remains highly relevant in the fields of information and communications, cybersecurity, and load optimization in information and communication systems. However, its detailed examination falls beyond the scope of this section.

Thus, this subsection provides an analysis of the *BMAP*-flow and its relationship with classical input flow models in DC-like systems. The *BMAP*-flow is modelled using an irreducible Markov chain, which governs the arrival of request batches with a certain probability, and its parameters are described through a system of square matrices. The outlined approach enables the formal determination of batch arrival intensities, state probabilities of the controlling Markov chain, and key flow characteristics. The methods for formal representation of the *BMAP*-flow through matrices A_n, $n \geq 0$, the generator G of transition probabilities, and their derived parameters have been examined. This approach enables the formalization of stochastic models of incoming flows in continuous time. The described mathematical tools allow for the estimation of key characteristics, including the average batch arrival intensity, variance of inter-batch intervals, and correlation coefficients between consecutive arrivals.

It should be noted that the dynamics of the *BMAP*-flow is determined by the controlling Markov chain ρ_t, $t \geq 0$, with a finite state space in continuous time. This characteristic imposes specific requirements on the approaches used to study processes, which are key to the analysis of DC-like systems. In particular, to examine the dynamics of the number of accepted requests in a service system at time t (denoted this process as j_t, $t \geq 0$), it is necessary to study the two-dimensional stochastic process (j_t, ρ_t), $t \geq 0$. This approach allows not only to account for the current state of the service system (the number of requests) but also to track changes in the internal states of the controlling Markov chain, which influence the nature of the arrival and processing of request batches. In the next subsection, the key results of research on two-dimensional Markov processes are discussed, which are especially useful for analysing DC-like systems with *BMAP*-flow.

2.3 Two-Dimensional Markov Chains as a Dynamic Component of the *BMAP*-Flow

Let us explore the two-dimensional Markov process $X_t = \{j_t, \vec{z}_t\}$, $t \geq 0$, with a state space $\Omega = \{(j, z), j \geq 0, z \in Z\}$ in continuous time, assuming that Z is a certain finite set. We can represent the generator P of the stochastic process $X_t = \{j_t, \vec{z}_t\}$, $t \geq 0$, in block form $P = \left(P_{j,m}\right)_{j,m\geq 0}$, where $P_{j,m}$ is the matrix formed by the intensities $p_{(j,\vec{y});(m,\vec{z})}$ of transitions of the Markov chain X_t, $t \geq 0$, from state $(j, \vec{y})$ to state $(m, \vec{z})$, $\vec{y}, \vec{z} \in Z$. The matrix $P_{j,j}$ contains diagonal elements of the form $p_{(j,\vec{y});(m,\vec{z})} = -\sum_{(i,\vec{y})\in Z\setminus(j,\vec{z})} p_{(j,\vec{y});(i,\vec{z})}$. It is important to note that a Markov process X_t, $t \geq 0$, is considered a quasi-birth-and-death process [173, 174] if its irreducible block generator $P = \left(P_{j,m}\right)_{j,m\geq 0}$ takes the form:

$$P = \begin{pmatrix} P_{0,0} & P_0 & \vec{0} & \vec{0} & \dots \\ P_2 & P_1 & P_0 & \vec{0} & \dots \\ \vec{0} & P_2 & P_1 & P_0 & \dots \\ \vdots & \vdots & \vdots & \vdots & \ddots \end{pmatrix}. \tag{2.3}$$

The existence of the stationary probability distribution of the Markov process X_t, $t \geq 0$, is ultimately ensured by the satisfaction of inequality:

$$\vec{w}P_2\vec{1} > \vec{w}P_0\vec{1}, \tag{2.4}$$

where the row vector $\vec{w}$ is the unique solution to the system of linear equations:

$$\vec{w}(P_0 + P_1 + P_2) = \vec{0}, \vec{w}\vec{1} = 1. \tag{2.5}$$

If the condition (2.4) is satisfied for the Markov process X_t, $t \geq 0$, the stationary probabilities of its states are determined by the limits $p(j, \vec{z}) = \lim_{t\to\infty} Q\{j_t = j, \vec{z}_t = \vec{z}\}, \vec{z} \in Z, j \geq 0$. These stationary probabilities are arranged in the row vector $\vec{p}_j = (p(j, \vec{z}), \vec{z} \in Z), j \geq 0$. The values of the stationary probabilities $\vec{p}_j, j \geq 0$, for the Markov process X_t, $t \geq 0$, are determined as

$$\vec{p}_j = \vec{p}_0\Lambda^j, j \geq 0, \tag{2.6}$$

where the matrix Λ is the unique positive solution with a spectral radius smaller than one for the matrix equation:

$$\Lambda^2 P_2 + \Lambda P_1 + P_0 = \vec{0}, \tag{2.7}$$

and the vector $\vec{p}_0$ is the unique solution to the system of linear equations:

$$\vec{p}_0 \left(P_{0,0} + \Lambda P_2\right) = \vec{0}, \underline{\vec{p}}_0 \vec{1} \Big/ (E - \Lambda) = 1. \tag{2.8}$$

A classical reference that describes the study of vector processes of birth and death is the fundamental article [175]. This work explores the key properties of such processes, including their stability, conditions for the existence of a stationary distribution, and analytical methods of investigation. Of particular interest is the problem of determining the stationary distribution of the waiting time for an accepted request for servicing in a DC-like system, whose behaviour is described by a generator of the form (2.3). To solve this problem, it is advisable to use the methods of studying Markov chains with absorbing states, which provide an effective approach to modelling the dynamics of such systems [94]. In particular, the use of results presented in [176] enables the derivation of analytical expressions for the probabilistic characteristics of the waiting process, which are critically important for performance analysis and optimization of service systems. The combination of methods for analysing vector processes of birth and death with the theory of Markov chains with absorbing states opens up prospects for the reliable description and prediction of the behaviour of DC-like systems operating under conditions of stochastic incoming request flows.

Now, let us examine the two-dimensional Markov process $X_k = \{j_k, \vec{z}_k\}$, $k \geq 1$, with a state space Ω in discrete time. The transition probability matrix for such a process is defined as $Q = \left(Q_{j,m}\right)_{j,m\geq 0}$, where the matrix $Q_{j,m}$ generalizes the probabilities $q_{(j,\vec{y});(m,\vec{z})}$ of one-step transitions of the Markov chain X_k, $k \geq 1$, from state $(j, \vec{y})$ to state $(m, \vec{z})$, $\vec{y}, \vec{z} \in Z$. It is appropriate to mention the one-step transition probability matrix of a multidimensional Markov chain of type $M|G|1$ [177]:

$$Q = \begin{pmatrix} B_0 & B_1 & B_2 & B_3 & \cdots \\ W_0 & W_1 & W_2 & W_3 & \cdots \\ \vec{0} & W_0 & W_1 & W_2 & \cdots \\ \vec{0} & \vec{0} & W_0 & W_1 & \cdots \\ \vdots & \vdots & \vdots & \vdots & \ddots \end{pmatrix}, \tag{2.9}$$

where block submatrices of type B represent transitions that increase the counting component of the Markov process, while block submatrices of type W characterize transitions that decrease the counting component or leave its value unchanged.

The quasi-Toeplitz transition probability matrix of the embedded Markov chain for a service system of type $M|G|1$, based on the moments of service completion, has a structure analogous to (2.9). In our case, the matrix (2.9) is a block upper Hessenberg matrix [178], where the j_k-component exhibits the skip-free to the left property. The corresponding generating functions have the form $W(s) = \sum_{j=0}^{\infty} W_j s^i$, $B(s) = \sum_{j=0}^{\infty} B_j s^i$, $|s| \leq 1$.

Assume that the Markov process $X_k = \{j_k, \vec{z}_k\}$, $k \geq 1$, is characterized by a transition probability matrix of the form (2.9), which is aperiodic and irreducible, and that the matrices $W(1)$ and $B(1)$ are stochastic and irreducible, satisfying the inequalities $W'(1) < \infty$, $B'(1) < \infty$. Then, for the Markov process X_k, $k \geq 1$, to be ergodic, it is necessary for inequality

$$[\det(sE - W(s))]'_{s=1} > 0 \tag{2.10}$$

to be satisfied. Note that for inequality (2.10), there exists an equivalent form:

$$\vec{u}W'(1)\vec{1} < 1, \tag{2.11}$$

where the vector $\vec{u}$ satisfies the system of linear equations:

$$\vec{u}W(1) = \vec{u}, \ \vec{u}\vec{1} = 1. \tag{2.12}$$

Let us assume that the condition (2.10) is satisfied and focus on studying the stationary probability distribution $p(j, \vec{z}) = \lim_{k\to\infty} Q\{j_k = j, \vec{z}_k = \vec{z}\}$, $\vec{z} \in Z$, of the Markov process $X_k = \{j_k, \vec{z}_k\}$, $k \geq 1$. We will arrange the stationary probabilities by introducing row vectors $\vec{p}_j = (p(j, \vec{z}), \vec{z} \in Z)$, $j \geq 0$, of length L. Let these vectors $\vec{p}_m$, $m \geq 0$, satisfy the Chapman system of equations:

$$\vec{p}_m = \vec{p}_0 B_m + \sum_{j=1}^{m+1} \vec{p}_j W_{m-j+1}, \ m \geq 0. \tag{2.13}$$

We recall the two most commonly used methods for calculating the vectors (2.13), namely, the generating function method and the matrix-analytic method.

Let us begin with the generating function method. Let the stationary distribution $\vec{p}_m$, $m \geq 0$, be fully characterized by the vector-generating function $\vec{\mathrm{P}}(s) = \sum_{m=0}^{\infty} \vec{p}_m s^m$. In this case, the function $\vec{\mathrm{P}}(s)$ is the solution (unique within the analytic circle $|s| < 1$ and continuous on the boundary $|s| = 1$) of the block functional equation:

$$\vec{\mathrm{P}}(s)(sE - W(s)) = \vec{p}_0(sB(s) - W(s)), \ |s| \leq 1. \tag{2.14}$$

If the behaviour of the studied DC-like system is governed by a multidimensional quasi-Toeplitz Markov process X_k, $k \geq 1$, then the functional Eq. (2.14) is the matrix analogue of the Pollaczek–Khinchine formula for the stationary distribution of the queue duration in a service system of type $M|G|1$. For such a model, with the service time distribution $D(t)$ and the intensity of the incoming request flow μ, we can write $W(s) = B(s) = d(\mu(1 - s))$, where $d(\omega) = \int_0^\infty \exp(-\omega t)dD(t)$, $\mathrm{Re}\,\omega \geq 0$. As a result, Expression (2.14) can be represented as $\mathrm{P}(s) = p_0 \frac{(s-1)d(\mu(1-s))}{s-d(\mu(1-s))}$.

Therefore, Expression (2.14) for a multidimensional Markov process coincides in form with its analogue for a one-dimensional Markov chain. For a service system of type $M|G|1$, the task of finding the stationary distribution essentially ends with obtaining the corresponding expression of the form (2.14), as the necessary probabilities p_0 for its calculation are determined by the normalization condition. In the case where the functional Eq. (2.14) is block-structured, the definition of the vector $\vec{p}_0$ becomes significantly more complex, but it is still an essential task.

In formalizing Expression (2.14), we considered the equilibrium Eq. (2.13). Thus, the determination of the vector $\vec{p}_0$ in the generating function method relies on the analyticity of the generating function $\vec{\mathrm{P}}(s)$ within the unit circle of the complex plane. Expanding on this statement, we note that the matrix $W(1)$ is irreducible, and among the elements of matrix $W(s)$, there is at least one element $W_{n,n'}(s)$ whose expansion $W_{n,n'}(s) = \sum_{m=0}^{\infty} W_m(n, n')s^m$ coefficients $W_m(n, n')$ in a series by powers satisfy the condition that there exists a $m, m \geq 0$, such that inequality $W_m(n, n')W_{m+1}(n, n') > 0$ holds. In this context, considering the ergodicity of the studied Markov process, we assume that equation

$$\det(sE - W(s)) = 0 \tag{2.15}$$

has exactly L multiple roots inside the circle $|s| \leq 1$, one of which is the root $s = 1$, and the remaining roots lie inside the unit circle. This statement is proven in [179, 180] with reference to Rouché's theorem.

Let the number of distinct roots s_n, $|s_n| < 1$, be $\tilde{L}$, and their multiplicity be characterized by the parameter k_n, $k_n \geq 1$, $n = \overline{1, \tilde{L}}$, $\sum_{n=1}^{\tilde{L}} k_n = L - 1$. Considering this, we can rewrite expression (2.14) as follows:

$$\vec{\mathrm{P}}(s) = \frac{\vec{\mathrm{P}}(0)(sB(s) - W(s))\mathrm{adj}(sE - W(s))}{\det(sE - W(s))}, \tag{2.16}$$

where $\mathrm{adj}(sE - W(s))$ defines the adjoint matrix for matrix $(sE - W(s))$. The adjoint matrix is calculated according to the rules for computing the inverse matrix, avoiding the final division operation.

Based on Expression (2.16), we can represent the vector $\vec{\mathrm{P}}(0)$ as part of the system of linear equations of the form:

$$\vec{\mathrm{P}}(0)\frac{d^k}{ds^k}\left[sB(s) - W(s)\mathrm{adj}(sE - W(s))\right]_{s=s_n} = 0, k = \overline{0, k_n - 1}, n = \overline{1, \tilde{L}}, \tag{2.17}$$

$$\vec{\mathrm{P}}(0)\frac{d}{ds}\left[sB(s) - W(s)\mathrm{adj}(sE - W(s))\right]_{s=1}\vec{1} = \frac{d}{ds}[\det(sE - W(s))]_{s=1}. \tag{2.18}$$

The system of linear Eqs. (2.17) and (2.18) has a unique solution $\vec{\mathrm{P}}(0) = \vec{p}_0$. Given the vector $\vec{\mathrm{P}}(0)$, the task of finding the stationary probability distribution of the Markov process X_k, $k > 0$, becomes trivial. Specifically, using Expression (2.14), we calculate the vector-generating function $\vec{\mathrm{P}}(s)$ and the factorial moments

$\vec{P}^{(l)} = \frac{d^l}{ds^l}\vec{P}(s)\Big|_{s=1}$, $l > 0$. It should be noted that the calculation of the moments $\vec{P}^{(l)}, l > 0$, is somewhat complicated by the fact that the matrix $(sE - W(s))$ at point $s = 1$ is degenerate.

Finally, to calculate an arbitrary number M of vectors $\vec{p}_m, m = \overline{0, M}$, the recursive formula

$$\vec{p}_{m+1} = \left(\vec{p}_m - \vec{p}_0 B_m - \sum_{i=1}^{m} \vec{p}_i W_{m-i+1}\right) \Big/ W_0, m \geq 0 \tag{2.19}$$

can be applied, defined in the basis of the equilibrium Eq. (2.13). It should be noted that in several cases, the calculation of the vectors $\vec{p}_{m+1}, m \geq 0$, using Expression (2.19) may be complicated by the degeneracy of the matrix W_0 or by the decrease in recursion stability due to the accumulation of rounding errors.

The generating function method is an effective tool for analysing Markov processes in $M|G|1$–type service systems and multidimensional quasi-Toeplitz Markov chains. This method is widely used in DC-like systems, network models, and stochastic processes, as it allows for obtaining compact analytical expressions for characteristic functions and solving equations in the form of matrix series. A significant contribution to the development of this approach was made by D. Iglehart, who proposed the matrix-analytic method [181–183] and thoroughly investigated the behaviour of Markov processes with batch arrivals of incoming requests. His work laid the foundation for further research in this area and greatly influenced the development of numerical methods for analysing service systems, particularly in reducing the computational complexity by adapting them to specific types of Markov processes. In particular, J. Abate and W. Whitt achieved significant results by focusing on the numerical aspects of implementing the generating function method in the context of information and communication systems, network traffic optimization, and the analysis of Markov processes with large state spaces.

An important element of the above-described generating function method is the procedure for finding the roots of the determinant of matrix $(sE - W(s))$ within the unit circle of the complex plane. In practice, the implementation of this procedure is often complicated by the potential lack of precision in the digit grid required for reliably separating the roots. This issue can be circumvented by the matrix-analytic method proposed by Neuts [184], which aims to determine the unknown vector $\vec{P}(0)$ without the need for finding the roots of the complex-variable function.

Let us consider the task of determining the vector $\vec{P}(0)$, where the ρ-th element $p(0, \rho)$ is inversely proportional to the average number of transitions the Markov chain performs after leaving state $(0, \rho)$ until it returns to $(0, \rho)$ for the first time. To achieve this, it is necessary to determine the distribution of the number of transitions of the Markov chain between two consecutive occurrences of states with a zero value of the counting component, simultaneously tracking the transitions of the finite component during this period. In this context, the concept of a fundamental period is introduced as the time interval from the moment when the value of the counting component equals h until the first moment when the value of this component reaches

$j-1, j>0$. By definition, for a Markov chain of type $M|G|1$, the duration of the fundamental period does not depend on j. Let us introduce the matrix generating function $H(s)$, where the (ρ, ρ')-th element contains the generating function of the number of transitions performed by the Markov chain during the fundamental period, while the finite component with the initial value ρ does not reach the value ρ': $\rho, \rho' = \overline{0, S}$. In the basis of the stochastic interpretation of the generating function in terms of "coloured" requests [185], the matrix $H(s)$ is defined as

$$H(s) = s\sum_{i=0}^{\infty} W_i H^i(s). \tag{2.20}$$

The dynamics of the counting component during the period between its consecutive entries into the zero state are as follows: on the first step, the counting component transitions from the zero state to some state $i \geq 0$, characterized by the transition probabilities of the finite component, generalized in the form of the matrix B_i. After that, if inequality $i \geq 1$ holds, $i \geq 1$ fundamental periods are realized, each accompanied by a sequential decrease of the counting component to zero with a step size of one.

Let us introduce the matrix generating function $L(s)$, where the (ρ, ρ')-th element contains the generating function of the number of transitions in the Markov chain from the moment it leaves the zero state with the finite component ρ until the counting component returns to the zero state with the finite component ρ': $\rho, \rho' = \overline{0, S}$. By analysing the dynamics of the counting component during the period between its consecutive entries into the zero state, we define the generating function $L(s)$ as

$$L(s) = s\sum_{i=0}^{\infty} B_i H^i(s). \tag{2.21}$$

We define the row vector $\vec{\xi}$ and the column vector $\vec{\xi}^*$ using the corresponding relations:

$$\vec{\xi}L(1) = \vec{\xi}, \vec{\xi}\vec{1} = 1, \vec{\xi}^* = L'(1)\vec{1}. \tag{2.22}$$

The element ξ_ρ of the vector $\vec{\xi}$ represents the stationary probability of the Markov process being in state ρ at the moment when the counting component exits the zero state: $\rho = \overline{0, S}$. The element ξ_ρ^* of the vector $\vec{\xi}^*$ represents the average number of transitions in the Markov chain during the period until the counting component returns to the zero state. At the moment of exiting the zero state, the finite component was in state $\rho = \overline{0, S}$.

Considering the preamble to Expressions (2.20) and (2.21) and relying on the principles of Markov renewal theory [186, 187], we express the desired definition of the vector $\vec{\mathrm{P}}(0)$ in terms of (2.22) as

$$\vec{\text{P}}(0) = \vec{\xi} \Big/ \vec{\xi}\vec{\xi}^{*}. \tag{2.23}$$

Expression (2.23) allows for the determination of the vector $\vec{\text{P}}(0)$ while avoiding the computation of the roots of Eq. (2.14). By extending the concept embedded in the matrix-analytic method, an alternative to the recursive procedure (2.19) can be derived for computing the vectors of stationary probabilities $\vec{p}_m$, $m > 0$. Below is a compact description of this adaptation:

1. By solving the nonlinear matrix Eq. (2.20) for $s = 1$ using the successive approximation method, we determine the generating matrix $H = H(1)$.
2. Compute the matrices $\overline{Q}_{i,j} = Q_{i,j} + \sum_{m=j+1}^{\infty} Q_{i,m} H^{m-j}$, $i \le j$.
3. Recursively compute the matrices Φ_m, $m \ge 0$: $\Phi_{m=0} = E$, $\Phi_{m>0} = \left(\overline{Q}_{o,m} + \sum_{i=1}^{m-1} \overline{Q}_{i,m}\Phi_i\right) \Big/ \left(E - \overline{Q}_{m,m}\right)$.
4. Find the unique solution $\vec{p}_0$ of the system of equations $\vec{p}_0\left(E - \overline{Q}_{0,0}\right) = 0$, $\vec{p}_0 \sum_{i=0}^{\infty} \Phi_i \vec{1} = 1$.
5. Determine the desired vectors $\vec{p}_i = \vec{p}_0 \Phi_i \partial x$, $i \ge 0$.

The advantage of this procedure over the one implemented in expression (2.19) is its computational stability, as all the matrices involved are positive.

Thus, this subsection focused on the representation of two-dimensional Markov chains, which define the dynamic component of the analytical modelling process in DC-like systems with *BMAP*-flows. The subsection outlined the mathematical structure of such flows, the conditions for the existence of their stationary distribution, and methods for its determination, including matrix-analytic approaches and the generating function method. It was demonstrated that the proper selection of parameters for the controlling Markov process allows for the optimization of load balancing in information and communication systems. The results presented in this subsection have practical applications in the modelling of broadband communication networks, optimization of flows in cloud computing, and load analysis in adaptive distributed systems, ensuring efficient resource management and enhanced service performance.

2.4 Single-Channel DC-Like Systems with *BMAP*-Flow

In this subsection, we examine single-channel DC-like systems with *BMAP*-flow, where the service time distribution follows a more generalized distribution than the classical Phase-Type (PT) distribution. The relevance of such systems is evident, as many real-world processes cannot be adequately modelled using only the *PH*-distribution, which imposes limitations on the accuracy of forecasting and the optimization of service system performance.

First, we will examine a DC-like system with an infinite buffer, where the service time distribution is arbitrary. This will allow us to investigate the general properties of such systems, assess their stability, and study the impact of the service time distribution type on the key performance indicators of this model. Next, we will consider a more complex system, where the service time is governed by a semi-Markov service process [188–190]. Such systems are crucial for modelling information and communication networks, adaptive control algorithms, and systems with stochastic delays, where service duration depends on external factors. We will then focus on DC-like systems with a finite buffer. These systems are critically important for modelling network traffic, resource-dependent information services, and distributed computing systems. The limitation on the maximum allowable queue length of accepted requests significantly affects operational efficiency and the probability of request loss in the analysed service system. Finally, we will examine retrial systems, where requests that fail to receive service immediately can retry to access the service facility after a certain time interval. Such models are widely used in the analysis of automated communication systems, cellular network queues, and for describing adaptive traffic routing. Thus, this subsection presents a comprehensive analysis of single-channel DC-like systems with a general service time distribution, enabling the assessment of their efficiency and the investigation of the impact of different queue types on the key performance characteristics of such systems.

A single-channel DC-like system with an infinite buffer, designed for servicing the incoming *BMAP*-flow, belongs to the type $BMAP|G|1$ [191]. As previously noted, the *BMAP*-flow is governed by a Markov process $\rho_t, t \geq 0$, with a state space $\{\overline{0, S}\}$ and a matrix generating function $A(s)$. The service durations of requests in a service system of type $BMAP|G|1$ are independent and follow the distribution function $D(t)$, with a Laplace–Stieltjes transform $d(\omega) = \int_0^\infty \exp(-\omega t) dD(t)$ and finite first moments $a_n = \int_0^\infty t^n dD(t), n = 1, 2$.

In the context of this section, an interesting topic is the analysis of the stationary behaviour of the stochastic processes j_t, which defines the number of accepted requests in the service system at moment t, and v_t, which characterizes the virtual waiting time in the system. Since these processes are non-Markovian, their stationary distribution cannot be analytically determined in explicit form using traditional methods. An effective approach to solving this problem involves the use of Markovization, which is implemented through the combined application of:

1. The state-space expansion method [192], which introduces an additional variable to model the controlling process.
2. The embedded Markov chain method [193], which focuses on the analysis of system states at service completion moments.

Let us assume that the k-th service completion moment of an accepted request is characterized by the variable t_k, $k > 0$, the number of accepted requests at moment t_k is j_{t_k}, and the state of the controlling Markov process at moment t_k is described by parameter ρ_{t_k}. From this set of parameters, we can conclude that the two-dimensional process $X_k = \{j_{t_k}, \rho_{t_k}\}$ forms a Markov chain. We organize the one-step transition probabilities of this chain in a structured manner $Q\{(j, \rho) \to (m, \rho')\} = Q\{j_{t_{n+1}} = m, \rho_{t_{n+1}} = \rho' | j_{t_n} = j, \rho_{t_n} = \rho\}, \rho, \rho' = \overline{0, S}$, and generalize them into square matrices $Q_{j,m}, j \geq 0, m \geq \max\{0, j-1\}$, of size $\overline{S} \times \overline{S}$.

The matrices $Q_{j,m}, j \geq 0, m \geq 0$, are defined in the form $Q_{o,m} = B_m, Q_{j,m} = W_{m-j+1}$, where $W_l = \int_0^\infty Q(l,t)dD(t), k \geq 0$. The procedures for determining the matrices $Q(l,t)$ through the matrices $L_l^{(i)}$ are described in Sect. 2.2.

$$B_l = \sum_{n=1}^{l+1} \int_0^\infty \exp(A_0 t)A_n dt W_{l+1-n} = -\sum_{n=1}^{l+1} A_n W_{l+1-n} \Big/ A_0. \tag{2.24}$$

Based on Expressions (2.24) and (2.9), the process X_k can be classified as a Markov chain of type $M|G|1$. Accordingly, the stationary distribution of the Markov chain X_k can be determined using the procedures described in Sect. 2.3. In particular, the generating functions $W(s) = \sum_{l=0}^\infty W_l s^l$, $B(s) = \sum_{l=0}^\infty B_l s^l$ will take the form $W(s) = -A(s)d = \int_0^\infty \exp(A(s)t)dD(t)$, $B(s) = -d(A(s) - A_0)A(s)\big/ sA_0$, respectively. Moreover, Expression (2.14), under the ergodicity condition $\beta = \mu a_1 < 1$, will take the form of the Pollaczek–Khinchine equation for a service system of type $M|G|1 : \vec{P}(s)(sE + A(s)d) = \vec{P}(0)A^2(s)d \Big/ A_0$.

It is important to note that the Markov process X_k, described above, is an auxiliary process, whereas the process $Y_t = \{j_t, \rho_t\}, j \geq 0$, holds greater practical significance. The process $\{j_t, \rho_t\}, t \geq 0$, is non-Markovian, but information about its distribution is embedded within the distribution of the Markov chain $\{j_{t_n}, \rho_{t_n}\}$. To extract this information, an approach based on embedded Markov renewal processes [194] is applied.

Let us assume that the probability distribution of the states of process $\{j_t, \rho_t\}$, $t \geq 0$, at moment t is defined as $q(j, \rho, t), j \geq 0, \rho = \overline{0, S}$, and its stationary distribution is characterized as $q(j, \rho), j \geq 0, \rho = \overline{0, S}$. We introduce the vectors $\vec{q}(j) = (q(j, 0), \ldots, q(j, S)), j \geq 0$, and their vector-generating function $\vec{q}(s) = \sum_{j=0}^\infty \vec{q}(j)s^j$. The generating function $\vec{q}(s)$ is expressed in terms of the generating function $\vec{P}(s)$ of the stationary distribution of the studied DC-like system at the moment of service completion for accepted requests:

$$\vec{q}(s)A(s) = \mu(s-1)\vec{P}(s). \tag{2.25}$$

We define the period that an accepted request would have to wait for service, after arriving at the service system at moment $t \geq 0$, as the virtual waiting period v_t. We introduce the row vector $\vec{S}(u)$, where the ρ-th element represents the stationary probability that, at a given moment, the controlling *BMAP*-flow Markov chain is in state $\rho \in \{\overline{0, S}\}$ and the virtual waiting period is less than u:

$$\vec{S}(u) = \lim_{t \to \infty} \{Q\{v_t < u, \rho_t = 0\}, \ldots, Q\{v_t < u, \rho_t = W\}\}. \tag{2.26}$$

We introduce the row vector $\vec{b}(\omega) = \int_0^\infty \exp(\omega u)d\vec{S}(u)$, formed by the results of the Laplace–Stieltjes transform of the elements of vector $\vec{S}(u)$, that is $\vec{b}(\omega)(\omega E + A(d(\omega))) = \omega\vec{q}(0)$, where $\mathrm{Re}\,\omega > 0$, $A(d(\omega)) = \sum_{n=0}^\infty A_n (d(\omega))^n$.

Let us assume that the actual waiting time for a request accepted at moment $t \geq 0$ for service is $v_t^{(r)}$. Then, the stationary probability that the actual waiting time is less than u is characterized as $S^{(r)}(u) = \lim_{t \to \infty} Q\left\{v_t^{(r)} < u\right\}$. The Laplace–Stieltjes transform of the distribution $S^{(r)}(u)$ is defined as $b^{(r)}(\omega) = \int_0^\infty \exp(-\omega u) dD^{(r)}(u)$, $\text{Re}\omega \geq 0$, or $b^{(r)}(\omega) = -\vec{b}(\omega)A(d(\omega))\vec{1} \Big/ \mu(1 - d(\omega))$. An arbitrary incoming request that arrives at the service system as part of a batch of n requests will, with probability $1/n$, be served as the i-th request within its batch, $i = \overline{1, n}$.

The step-by-step derivation of the individual results mentioned above is presented in the works of Chakravarthy [195], as well as in earlier studies by Ferng [196]. A comprehensive justification of these results, supplemented with rigorous mathematical proofs and significant generalizations, can be found in [197].

Among the alternative sources covering some of the aspects discussed, it is worth mentioning the chapter “Probability Theory and Mathematical Statistics” [198] and the book “Selected Works II. Probability Theory and Mathematical Statistics” [199]. In [200], using the method of introducing an additional variable, the probability distribution of the number of accepted requests in the service system at an arbitrary moment in time was obtained. The study [201] is dedicated to proving the validity of formula (2.25), which establishes the relationship between the generating functions of the stationary distribution of service system states at an arbitrary moment in time and at service completion moments. This dependence is examined for a wide class of DC-like systems with *BMAP*-flow, but the numerical implementation of the corresponding recursive algorithms is potentially unstable, making the accurate computation of the stationary distribution challenging and requiring expert intervention. It should be noted that the source of numerical instability in the recursion obtained by expanding Eq. (2.25) into a Taylor series, which is used to compute the stationary state probabilities of the studied service system at an arbitrary moment through the probabilities of the embedded Markov chain, is the matrix subtraction operation. However, in [202], a stable recursion for computing the mentioned probabilities was proposed.

In [203], a factorization method is presented, aimed at determining the distribution of the number of accepted requests in the service system and their waiting time for service. It is worth noting that, in certain cases, the use of special classes of service time distributions can significantly simplify the analysis. For example, this approach was applied in [204] for a service system of type $MAP|D_N|1$, where the service duration is randomly selected from N predetermined values.

Similar results have been obtained for DC-like systems with deterministic service duration [205]. In [206], the case where the Laplace–Stieltjes transform of the service time distribution has a fractional-rational form is investigated. In [207], a service system of type $BMAP|MSP|1$ is considered, where the service process is Markovian and structurally similar to the Markovian input flow. The study of an DC-like system of type $BDMMAP|G|1$, in which the controlling process of the *BMAP*-flow is described not by an arbitrary continuous-time Markov chain, but by a birth-and-death process, is presented in [208]. This approach allows for better modelling of queue dynamics in cases where the structure of the input flow cannot be described using standard *BMAP*-models. Finally, in [209], an analysis of a service system of type $BMAP|G|1$ is conducted for the LIFO (last-in-first-out) service discipline with probabilistic priority. The studies mentioned in this paragraph demonstrate the diversity of mathematical methods used for analysing Markov models of DC-like systems with *BMAP*-flows. The use of recursive approaches, factorization methods, specific classes of service time distributions, and alternative models of controlling processes, such as birth-and-death processes, significantly expands the capabilities of stochastic modelling for complex service systems.

The *BMAP|G|1*–type model mentioned above deserves special attention in the context of our study. Such a DC-like system is based on the assumption that service durations of requests are independent stochastic variables with a single distribution and distribution function $D(t)$. However, in many real-world service systems, consecutive requests exhibit stochastic dependence, and their service durations may be determined by different distribution laws. To model such cases, the semi-Markov (SM) process formalism has been proposed in the literature [33, 34, 188–190]. Let us outline the key aspects of this approach.

Let us assume the existence of a stochastic semi-Markov process l_t, $t \geq 0$, with a stationary probability distribution. Such a process is fully defined by its semi-Markov kernel and state space. We define the kernel of the process as the matrix $\vec{D}(t)$, composed of elements $D_{l,l'}(t)$, $l, l' = \overline{1, Z}$, while the state space is characterized by the set $\{1, 2, \ldots, Z\}$. The function $D_{l,l'}(t)$ represents the probability that the duration of the process l_t, $t \geq 0$, in its current state l is less than t, and that the next state after the transition will be l'.

The service process will be semi-Markovian if the service durations of consecutive requests are characterized by sequential periods of the process l_t, $t \geq 0$, remaining in its respective states. Let us analyse the behaviour of a DC-like system with a semi-Markovian controlling process or a service system of type $BMAP|SM|1$ within the generating function formalism.

First, let us examine the embedded process $\{j_k, \rho_k, l_k\}$, where

– j_k is the number of accepted requests at moment t_k, which marks the completion of service for the k-th request;
– ρ_k is the state of the controlling *BMAP*-flow process at moment t_k;
– l_k is the state of the *SM*–type controlling service process at moment t_k.

It is evident that the process l_t, $t \geq 0$, defined in this way, forms a three-dimensional quasi-Toeplitz Markov chain.

In our analysis of DC-like systems of type $BMAP|G|1$, we have already described a two-dimensional quasi-Toeplitz Markov chain. In this case, the components $\{\rho_k, l_k\}$ have a finite set of values, which allows them to be formally replaced by an aggregating component, enabling the three-dimensional process l_t, $t \geq 0$, to be analysed as a two-dimensional one. However, we will not implement this simplification directly. Instead, we will analyse the original *BMAP|SM|1*–type model, employing Kronecker products and matrix summations.

The Kronecker product A ⊗ B of matrices $A = (\alpha_{i,j})$ and $B = (\beta_{i,j})$ results in a matrix composed of blocks $(\alpha_{i,j}B)$. The Kronecker sum A ⊕ B of matrices $A = (\alpha_{i,j})$ and $B = (\beta_{i,j})$ is the sum of matrices $A \otimes E_B + E_A \otimes B$, where E_A and E_B are identity matrices of the same dimensions as matrices A and B, respectively. In the context of our study, the usefulness of these operations becomes apparent when we interpret matrices A and B as the one-step transition probability matrices of independent Markov chains $X_k^{(1)}$ and $X_k^{(2)}$. Then, the matrix A ⊗ B represents the transition probability matrix of the two-dimensional Markov chain $\left\{X_k^{(1)}, X_k^{(2)}\right\}$. Next, we will also use the mixed product rule, analytically defined as $(A\Gamma)\otimes(B\Delta) = (A \otimes B)(\Gamma \otimes \Delta)$. Based on these matrix operations, the generating functions $W(s)$ and $B(s)$, which characterize the one-step transition probabilities of the Markov chain $\{j_k, \rho_k, l_k\}$, $k > 0$, can be represented as

$$W(s) = \int_0^\infty \exp(A(s)t) \otimes d\vec{D}(t) = -\hat{d}A(s),$$

$$B(s) = \left(\tilde{A}(s) - \tilde{A}_0\right)\hat{d}A(s) / s\tilde{A}_0,$$

where $\tilde{A}_n = A_n \otimes E_Z$, $\tilde{A}(s) = A(s) \otimes E_Z$. It is evident that while these matrix generating functions $W(s)$ and $B(s)$ are essentially analogous to those used for a $BMAP|G|1$ -type service system, the integrand matrix multiplication operation is replaced by the Kronecker product operation.

A necessary and sufficient condition for the existence of a stationary distribution in the Markov chain $\{j_k, \rho_k, l_k\}$, $k > 0$, is the satisfaction of inequality:

$$\beta = \mu a_1 < 1, \tag{2.27}$$

where $\mu = \vec{\pi}A'(1)\vec{1}$ is the average arrival rate of requests in the *BMAP*-flow; $a_1 = \vec{\gamma}\int_0^\infty t d\vec{D}(t)\vec{1}$ is the average service time; and $\vec{\gamma}$ is the invariant vector of the stochastic matrix $\vec{D}(\infty)$, which satisfies the system of equations $\vec{\gamma}\vec{D}(\infty) = \vec{\gamma}$, $\vec{\gamma}\vec{1} = 1$. From this point forward, we assume that condition (2.27) holds, and thus the stationary probabilities $p(j, \rho, l) = \lim_{k\to\infty} Q\{j_k = j, \rho_k = \rho, l_k = l\}, j \geq 0, \rho = \overline{0, S}$, $l = \overline{1, Z}$, exist.

Let $\vec{p}(j, \rho) = (p(j, \rho, 1), \ldots, p(j, \rho, Z))$, $\vec{p}_j = (p(j, 0), \ldots, p(j, S))$, $\vec{P}(s) = \sum_{i=0}^\infty \vec{p}_i s^i$, $|s| < 1$, be defined. The vector-generating function $\vec{P}(s)$ of the stationary distribution of the embedded Markov process $\{j_k, \rho_k, l_k\}$, $k > 0$, satisfies the matrix functional equation $\vec{P}(s)\left(-\hat{d}A(s) - sE\right) = -\vec{P}(0)\tilde{A}(s)\hat{d}A(s) \Big/ \tilde{A}_0$. To compute the vector $\vec{P}(0)$ and the generating function $\vec{P}(s)$, given the known matrix generating functions $W(s)$ and $B(s)$, we can use the approaches previously mentioned in our analysis of the DC-like system of type $BMAP|G|1$.

If condition (2.27) holds, then the stationary distribution $q(j,\rho,l) = \lim_{t\to\infty} Q\{j_t = j, \rho_t = \rho, l_t = l\}$ for the state probabilities of the semi-Markov process $\{j_k, \rho_k, l_k\}$, $k > 0$, also exists. We introduce the notation

$$\vec{q}(j,\rho) = (q(j,\rho,1),\ldots,q(j,\rho,Z)),$$

$$\vec{q}(j) = (\vec{q}(j,0),\ldots,\vec{q}(j,S)), \vec{\mathrm{Q}}(s) = \sum_{i=0}^{\infty}\vec{q}(i)s^i, |s| < 1,$$

where $\vec{\mathrm{Q}}(s)$ is the generating function of the stationary distribution of process $\{j_k, \rho_k, l_k\}$, $k > 0$, at an arbitrary moment in time. The generating functions $\vec{\mathrm{Q}}(s)$ and $\vec{\mathrm{P}}(s)$ at the moment of service completion for accepted requests are related as follows:

$$\vec{\mathrm{Q}}(s)\tilde{A}(s) = -\mu\vec{\mathrm{P}}(s)\left(\left(s\nabla^*(s)/\hat{d}A(s)\right) - E\right),$$

where $\nabla^*(s) = \int_0^\infty \exp(A(s)t) \otimes d\nabla_D(t)$ is a diagonal matrix with nonzero elements $\left(\vec{D}(t)\vec{1}\right)_i$, $i = \overline{1,Z}$. The expressions for computing the Laplace–Stieltjes transform of the virtual and actual waiting times of an accepted request for service are similar to the corresponding expressions given earlier in the analysis of the DC-like system of type $BMAP|G|1$ (see Expression (2.26) and further).

In addition to the fundamental work [210], among the studies dedicated to the analysis of DC-like systems of type $BMAP|SM|1$, the work of F. Machihara [211] is particularly noteworthy. This study examines an extended model that considers the stochastic dependence between the arrival and service processes. Such an approach allows for a more realistic representation of the behaviour of information and communication systems, computer networks, and multi-channel information services, where request flow parameters may vary depending on the load of the servicing stations. Unlike the classical $BMAP|SM|1$–type model, which assumes independence between flows, the study [212] proposes a new method for determining the correlation dependency between the incoming request flow and the service durations of accepted requests. To achieve this, the author applied a modified Markovization approach, preserving the structural property of quasi-Toeplitz Markov chains. The results of this study expand the analytical framework for modelling DC-like systems of type $BMAP|SM|1$, which can be valuable for optimizing service system performance, predicting delays, and balancing workloads.

An extension of the $BMAP|SM|1$–type model concept is service systems where the input buffer has a finite capacity N, which defines the maximum number of accepted requests that can wait for service. Finite buffer capacity is a key feature of most real-world systems, particularly in information and communication networks, distributed computing, and information services. This makes the study of such models a relevant and important area of applied mathematics.

A significant contribution to the study of DC-like systems with a finite buffer and *MAP*-input flow (as well as its special case—the *PH*-flow) is represented in [213]. Additionally, key aspects of such models were explored in earlier studies [214, 215]. The *MAP*-flow is an ordinary analogue of the *BMAP*-flow, allowing for simplified analysis in certain cases. However, in single-channel service systems with an arbitrary service time distribution and an unlimited buffer, using the *MAP*-model instead of the *BMAP*-model does not provide significant computational advantages. This is because both approaches require the stationary distribution analysis of a two-dimensional quasi-Toeplitz Markov process, governed by matrix-block transition probabilities. In contrast, for service systems where the service duration follows a phase-type distribution (*PH*-distribution), eliminating group arrivals significantly simplifies the analysis. In this case, it is sufficient to determine the stationary distribution of the vector birth-and-death process, which is defined by a generating matrix of the corresponding structure.

The study of DC-like systems with a finite buffer and *BMAP*-flow presents additional analytical challenges due to the potential conflict between group arrivals of requests and the finite buffer capacity. In particular, there may be situations where buffer space is available, but not enough to accommodate the entire group of incoming requests. In such cases, it is necessary to determine which requests will be accepted and which will be rejected. To formalize these situations, the literature considers three main disciplines for managing incoming request flows, which define how the model behaves in cases of partial buffer occupancy:

1. Partial Admission (PA)—only those incoming requests that fit into the buffer are added to the queue, while the remaining requests are lost.
2. Complete Admission (CA)—the entire group of requests is accepted, even if this temporarily expands the buffer capacity.
3. Complete Rejection (CR)—if there is not enough space in the service system to accommodate the entire group of incoming requests, none of the requests are accepted, and the entire group is rejected.

The choice of a specific discipline depends on the application characteristics of the DC-like system, particularly on whether the requests within a group are independent or represent a single informational entity, as well as on the availability of reserved buffer space or memory management policies. Among the approaches considered, the *PA*-discipline is the most extensively studied, as it provides an optimal compromise between efficient buffer utilization and minimization of incoming request losses.

The estimation of the probability of incoming request loss in an DC-like system of type *BMAP*|*G*|1|*N*, based on the approximation of the stationary distribution of the corresponding model with an unlimited buffer, was first proposed in [86]. Subsequent research [216] developed a computationally efficient but unstable method for calculating the stationary state probabilities for the *BMAP*|*G*|1|*N*–type model with *PA*-discipline. This study also introduced an adaptive hysteresis dependence, where the operating mode of the service system is influenced by the number of accepted requests. Further advancements in this research direction were presented in [217], where a stable algorithm for computing stationary probabilities in a DC-like system of type $BMAP|G|1|N$ with *PA*-discipline was proposed. The mentioned algorithm is based on sensing the output state of the Markov chain, which allows transforming the corresponding equilibrium equation system into a numerically stable form. A similar approach was applied in [218] to analyse *CA*– and *CR*–disciplines, comparing the efficiency of these group request acceptance policies in service systems. The study [219] introduced a randomized strategy for group request acceptance, which involves the stochastic application of one of the *PA*–, *CA*– or *CR*–disciplines at the moment of request arrival. This approach enables modelling adaptive systems that automatically adjust queue management policies for accepted requests based on their current state. Meanwhile, [220] analysed an $BMAP|G|1|N$–type model with *PA*- and *CA*-disciplines, where the service unit enters a "vacation" mode once the queue is emptied. This mechanism helps reduce system load and improve resource efficiency. As part of the analysis of non-stationary modes, study [221] examined a DC-like system of type $MAP|G|1|N$, deriving the distribution of the time until the first buffer overflow, which is a critical metric for real-world information systems. A similar approach was applied in [222] to analyse an $BMAP|G|1|N$–type model, while [164] studied a $MAP|G|1|N$–type model under conditions where the coefficient of variation of service duration is high, which is typical for networks with a high degree of traffic stochasticity.

The analysis of the mentioned studies leads to the conclusion that incorporating *BMAP*-flow into the formalization of service system operations with a finite buffer introduces specific challenges related to group request arrivals and potential losses due to resource limitations. As a result, the development of computationally stable methods for calculating stationary probabilities remains a key task in obtaining reliable analytical estimates of such models' characteristics. Future research in this area could focus on developing adaptive queue management strategies, based on the dynamic selection of group request processing disciplines. Additionally, a promising direction is the numerical analysis of service systems with variable priorities and dynamic buffer capacities, which is particularly relevant in the context of resource optimization in distributed information and communication systems, where effective queue management is a critical factor for performance.

So far, we have analysed DC-like systems with either an unlimited or a finite buffer. However, apart from these models, there exists a distinct class of service systems with retrial calls, which is characterized by the absence of a buffer for accepted requests. In such service systems, a request arriving when the service unit is busy is not lost entirely but instead enters a waiting state on the caller's side. After a certain period, it attempts again to be accepted for service. If, during the retrial attempt, the service unit is still occupied, the request continues making new attempts until one is successful. Requests undergoing retrial attempts are considered to be "in orbit". This behavioural model is typical for many information and communication systems, including cellular networks, call management systems, and distributed computing environments, where failed requests are automatically retried via automatic redial mechanisms.

The study of DC-like systems with retrial calls has attracted significant attention from researchers since the mid-twentieth century. The state of research in this field has been discussed in numerous survey papers, among which [223, 224] are particularly noteworthy. A major contribution to the theoretical foundations of this field was made by G. Falin, who studied $M|G|1$ – and $M|M|N$–type models with retrial calls and the classical redial strategy. For such service systems, a characteristic pattern emerges: if i requests are in orbit, the total retrial intensity α_i equals $i\alpha$, where α is the individual retrial intensity for a single request. This model realistically represents the redistribution of load under high resource competition conditions. The most comprehensive presentation of the analysis of DC-like systems with retrial calls is provided in the survey [225]. In particular, let us consider the embedded Markov chain at service completion moments in an $M|G|1$ -type service system with retrial calls. The generating function of the stationary distribution of this Markov process can be characterized by the differential equation:

$$p'(s) = p(s)g(s), \tag{2.28}$$

where the function $g(s) = \frac{1}{\phi}(s)\phi'(s) - \frac{\mu}{\alpha\phi}\phi(s)\left(\frac{1}{\phi(s)} - 1\right)$ defines the Pollaczek–Khinchine formula $p(s) = p(0)\phi(s)$ for the corresponding model of type $M|G|1$ with an infinite buffer.

The service system with retrial calls and *BMAP*-flow was explicitly studied in [226]. This study presents the differential form of the vector-generating function $\vec{\mathrm{P}}(s)$ for the stationary distribution of service completion moments in the form of the equation:

$$\vec{\mathrm{P}}'(s) = \vec{\mathrm{P}}(s)\vec{\mathrm{G}}(s),$$

where $\vec{\mathrm{G}}(s) = \vec{\Phi}'(s) \Big/ \vec{\Phi}(s) + \left(\vec{\Phi}(s)\left(1 / \Phi(s) - E\right)\right) \Big/ s\alpha A_0$ is the vector-generating function and $\vec{\Phi}(s) = -\frac{A(s)d(-A(s))}{A_0(sE-d(d(-A(s))))}$ is a matrix function, which serves as an analogue of the Pollaczek–Khinchine formula $\vec{\mathrm{P}}(s) = \vec{\mathrm{P}}(0)\vec{\Phi}(s)$ for the *BMAP*|*G*|1–type model with an infinite buffer.

The results described by Eq. (2.28) and $\vec{\mathrm{P}}'(s) = \vec{\mathrm{P}}(s)\vec{\mathrm{G}}(s)$, obtained for *M*|G|1– and *BMAP*|*G*|1–type service systems with retrial calls, share a similar mathematical structure. However, the linear homogeneous differential Eq. (2.28) can be straightforwardly solved, considering the initial condition $p(0) = 1 - \beta$, which directly follows from the normalization relation $p(1) = 1$. In contrast, equation $\vec{\mathrm{P}}'(s) = \vec{\mathrm{P}}(s)\vec{\mathrm{G}}(s)$ lacks an analytical solution. In the context of $\vec{\mathrm{P}}'(s)$, a direct generalization of the matrix-form solution for Eq. (2.28), expressed as an exponential dependence, is potentially incorrect, since the matrix $\vec{\mathrm{G}}(s)$ may not commute with the derivative $\int \vec{\mathrm{G}}(s)ds$. An additional complicating factor is the unknown values of the initiating vector function $\vec{\mathrm{P}}(0)$, which prevents the direct application of standard integration methods. Nevertheless, equation $\vec{\mathrm{P}}'(s) = \vec{\mathrm{P}}(s)\vec{\mathrm{G}}(s)$ can be effectively used to derive relationships between the factorial moments of the distribution of accepted requests in a DC-like system of type $BMAP|G|1$ with retrial calls. In study [227], the results of [226] were generalized to a *BMAP|G|1*–type model with retrial calls. Additionally, the work examined a linear retrial strategy, where the total intensity of retrial attempts is determined by equation $\alpha_i = i\alpha + \varphi$, $\alpha, \varphi \geq 0$. A special case of this approach, where the intensity of retrial attempts remains constant and does not depend on the number of requests in orbit, is particularly relevant when modelling service systems with centralized retrial management. In such models, it is assumed that either: there is a queue in orbit, and only the first request in the queue is allowed to generate retrial attempts; retrial attempts occur independently, but their intensity decreases as the number of requests in orbit increases. In the same study [226], an algorithm was proposed for determining the unknown vector $\vec{P}(0)$, which serves as the initial condition for the stationary distribution equation. Additionally, a relationship was established between the state distributions of the service system at embedded and arbitrary time points, which is analogous to the one formulated for an *BMAP|SM|1*–type model with an unlimited buffer. In study [227], a DC-like system of type $MAP|PH|1$ with retrial calls was examined, where the service time is modelled by a phase-type distribution. This generalization enables the modelling of systems where the service duration follows a multiphase structure, providing a more accurate approximation of complex stochastic processes. The analysis of such dependencies between service phases will be discussed in the next subsection.

Study [228] extends the results of [225] to an *BMAP|SM|1*–like retrial system, where the redial intensities α and φ, which define the linear retrial strategy, are not deterministic but instead stochastically vary based on the state of an external Markov process with a finite state space. In study [229], a DC-like system of type $BMAP|SM|1$ was investigated, in which the operational characteristics vary depending on the state of the governing Markov random process. This model can operate in various modes, including a service system with an unlimited buffer, a bufferless service system, a classical retrial system, or a retrial system with a fixed redial intensity. Study [230] focuses on an *BMAP|G|1*–type retrial system, which, in addition to the standard retrial mechanism from the orbit, introduces a proactive search by the idle service station for requests in the orbit. This search occurs over a stochastically distributed time period with an arbitrary distribution. This model represents scenarios where elements of dynamic communication networks or autonomous queue management systems actively initiate requests from the orbit, rather than waiting for redials.

Since the generating function method proved ineffective for analysing DC-like systems with *BMAP*-flow and retrials, an alternative approach was formulated for studying such models. The main difficulty in analysing these DC-like systems, compared to unbuffered service system models, lies in the following: in service systems with an unlimited queue, the behaviour of the embedded Markov chain $\{j_k, \rho_k\}$, $k > 0$, is independent of the value $j > 0$. This ensures a quasi-Toeplitz structure of the block matrix of one-step transition probabilities, which significantly simplifies the analysis of the model's dynamics. In turn, in service systems with repeated calls, the behaviour of the embedded Markov chain $\{j_k, \rho_k\}, k > 0$, depends on the value of $j > 0$. Investigating such a spatially inhomogeneous Markov process is more challenging. However, for large values of the first component of the chain, this dependence weakens for the following reason: when the number of requests in the orbit is large, the service station is almost never idle and, upon completion of service, is immediately seized by a new request from the orbit. This creates an effect analogous to the operation of a service system with an unlimited buffer, in which the service station is always engaged in processing the next accepted request from the buffer. Consequently, the process of changes in the number of requests in the orbit acquires the characteristics of a queue in a service system with an unlimited buffer. These considerations formed the basis for introducing a new class of Markov processes, known as multidimensional asymptotically quasi-Toeplitz Markov chains [231]. In addition to defining this class of Markov processes, the aforementioned studies also formulated the necessary and sufficient conditions for the ergodicity of such Markov chains and presented an algorithm for determining their stationary distribution for both discrete- and continuous-time cases.

2.5 Multi-channel DC-Like Systems with *BMAP*-Flow

It is well known that, in the general case, an exact analytical analysis of multi-channel service systems with an arbitrary service time distribution is impossible, even if the incoming flow is a stationary Poisson process (except for models with an unlimited number of channels or models with request losses). Given this circumstance, the study of multi-channel DC-like systems with *BMAP*-flow relies on the most general representation of the service time distribution (typically, choosing a phase-type distribution). This approach is discussed below.

Introduce a stochastic vector $\vec{d} = (d_1, \ldots, d_z)$ of length Z and a subgenerator Ω of order Z in the form of a square matrix of the corresponding size, composed of negative diagonal elements $\Omega_{l,l}$ and positive elements $\Omega_{l,l'}$, $l' = l, l = \overline{1, Z}$, for which $\sum_{l'=1}^{Z} \Omega_{l,l'} \leq 0 \forall l$ and at least one l satisfy the condition $\sum_{l'=1}^{Z} \Omega_{l,l'} < 0$. Define a stochastic variable X with a phase-type distribution, represented in an irreducible form $(\vec{d}, \Omega)$ (with an irreducible matrix $\Omega + \vec{\Omega}_0\vec{d}$). Let there exist a continuous-time Markov chain $\lambda_t, t \geq 0$, with a state space $\{1, \ldots, Z, Z+1\}$, where state $Z + 1$ is the only absorbing state. At the initial moment $t = 0$, the Markov process $\lambda_t, t \geq 0$, assumes the value l, $l \in (1, \ldots, Z)$, with probability d_l. The duration of the Markov chain $\lambda_t, t \geq 0$, remaining in state l is characterized by an exponential distribution with parameter $-\Omega_{l,l}$. Subsequently, this process transitions to state l', $l' \in (1, \ldots, Z)$, $l' \neq l$, with intensity $\Omega_{l,l'}$, whereas the transition intensity of process $\lambda_t, t \geq 0$, from state l to the absorbing state $Z + 1$ is characterized by probability $-\sum_{l'=1}^{Z} \Omega_{l,l'}$. The value of the stochastic variable X with a phase-type distribution is exhausted at the moment when the Markov process $\lambda_t, t \geq 0$, transitions to the absorbing state. Introduce a positive vector $\vec{\Omega}_0 = -\Omega\vec{1}$ with at least one nonzero element. Any arbitrary l-th element $\left(\vec{\Omega}_0\right)_l$ of vector $\vec{\Omega}_0$ characterizes the probability of process $\lambda_t, t \geq 0$, transitioning to the absorbing state from state $l, l \in (1, \ldots, Z)$.

Interpret time as a stochastic parameter with a phase-type distribution in its continuous representation $(\vec{d}, \Omega)$. A virtual incoming request enters a network consisting of Z nodes and performs inter-node transitions within it. The first node is selected arbitrarily from the set $\{1, \ldots, Z\}$ according to the distribution specified by the components of vector $\vec{d}$. Upon reaching the l-th node, the virtual request remains there for a period characterized by a stochastic variable following an exponential distribution with parameter $-\Omega_{l,l}$. Upon the completion of this period, the virtual request transitions to node $l' = \overline{1, Z}, l' \neq l$, with probability $-\Omega_{l,l'}/\Omega_{l,l}$ or to the absorbing state with probability $1 + \sum_{l'=1,l'\neq l}^{z} \frac{\Omega_{l,l'}}{\Omega_{l,l}}$. Thus, the time until the virtual request reaches the absorbing state is a stochastic variable u with a phase-type distribution, represented irreducibly by $(\vec{d}, \Omega)$ and characterized by the distribution function $\mathrm{A}(u) = 1 - \vec{d} \exp(\Omega_u)\vec{1}$. If the irreducible representation $(\vec{d}, \Omega)$ is defined, then determining the distribution function $\mathrm{A}(u)$ becomes a trivial task. At the same time, the inverse problem does not have a unique solution. The Laplace–Stieltjes transform $\alpha(\omega) = \int_0^\infty \exp(-\omega t) d\mathrm{A}(t)$, $\cdot \mathrm{Re}\omega \geq 0$, of the phase-type distribution is defined as $\alpha(\omega) = \vec{d}\vec{\Omega}_0/(\omega E - \Omega)$, while the l-th-order initial moment of this distribution is given by $a_l = \int_0^\infty t^l d\mathrm{A}(t) = l!\vec{d}\,\overrightarrow{1}/(-\Omega)^l$.

It is known [135] that for any distribution function $\mathrm{A}(t)$ of a positive stochastic variable, a phase-type distribution can be selected that approximates $\mathrm{A}(t)$ arbitrarily closely in the sense of weak convergence. In the study of DC-like systems of type $BMAP|G|1$ presented in the previous subsection, the matrix $W_l = \int_0^\infty Q(k, t) dD(t)$, $k \geq 0$, was used, where the matrix $Q(k, t)$ was expressed through recursively computed estimators of matrix $L_k^{(i)}$ (see Sect. 2.3). If the distribution $D(t)$ is of the phase type, then the matrices W_m, $m \geq 0$, defined by the irreducible representation $(\vec{d}, \Omega)$, are computed using a much simpler expression $W_m = C_m\left(E_{\overline{S}} \otimes \vec{\Omega}_0\right)$, where the matrices C_m, $m \geq 0$, are determined from the recursion $C_0 = -\left(E_{\overline{S}} \otimes \vec{d}\right)/(A_0 \oplus \Omega)$, $C_m = -\sum_{i=0}^{m-1} \frac{C_i(A_{m-i} \otimes E_Z)}{(A_0 \oplus \Omega)}$, $m > 0$.

One of the first studies dedicated to multi-channel DC-like systems with *BMAP*-flow and an exponential service time distribution is [232]. This work analyses service systems based on vector-generating functions and employs known properties of multidimensional quasi-Toeplitz Markov processes. Particular attention is given to the possibility of solving a functional equation of type (2.14), which describes the stationary characteristics of the corresponding service systems.

The study of service processes with a phase-type distribution in the context of multi-channel service systems does not fundamentally complicate the analytical constructions compared to models where the service time follows an exponential distribution. However, such a modification significantly increases the technical complexity of calculations and complicates the computational implementation of numerical algorithms used for the analysis of the examined service system instance.

If the examined DC-like system is multi-channel, the Markov chain describing its dynamics must account not only for the current number of accepted requests in the model and the state of the controlling *BMAP*-flow process but also for the configuration of the states of the service control processes at each service station. To reduce the model's dimensionality, a numbering of occupied channels can be introduced, allowing the transformation of the original multidimensional Markov chain into a two-dimensional one. This approach enables the analysis of the service system to be reduced to the study of a quasi-Toeplitz Markov process with discrete time by generalizing the material from Sect. 2.2, where a similar analysis of quasi-Toeplitz Markov chains with continuous time was demonstrated. In the case of *MAP*-flow, analytical investigation is significantly simplified, as the dynamics of the DC-like system can be represented as a vector branching and extinction process. Moreover, for most states (except for a few boundary cases) the stationary probability distribution acquires a matrix-analytic structure, which considerably simplifies both analytical support and the numerical implementation of the corresponding algorithms.

Let us proceed with the study of the DC-like system of type $BMAP|PH|N|0$. The relevance of such service systems is driven by several factors. Firstly, this model represents a significant modification of the classical multi-channel *M*I*M*I*N*I0–type model, which marked the beginning of the development of queueing theory. Notably, it was for the $M|M|N|0$ model that the *B*-Erlang formula was derived, which for a long time served as the primary tool for designing telephone networks. Secondly, it is well known that the state probability distribution for an *M*I*M*I*N*I0–type service system coincides with the corresponding distribution for a $M|G|N|0$ model, even if the service time distributions are arbitrary (provided they have the same mean values). Accordingly, the question arises as to whether this property holds when the incoming flow is not a stationary Poisson process but rather a *BMAP*– or *MAP*–flow. To answer this, recall the unique property of a stationary Poisson flow, known as individual randomness: if it is known that k events have occurred within an interval $[0, t]$, then their arrival times are independent of each other and uniformly distributed over $[0, t]$. Since the *BMAP*-flow does not possess a similar property, it can be expected that the invariance of the distribution, proven for Poisson flows, does not hold for the *BMAP*-flow. In the study [223], an analysis of the *BMAP*I*PH*I*N*I0–type model was conducted for all the aforementioned service system access disciplines (*PA*–, *CA*–, and *CR*–disciplines). The stationary state probability distribution was obtained, and the probability of losing an arbitrary incoming request was determined. Additionally, an extensive numerical experiment was performed, the results of which demonstrated a significant impact on the probability of request loss. This impact was influenced both by the correlation properties of the incoming flow (under the same average request arrival rate) and by the coefficient of variation of the service time distribution (for a fixed expected service time). The obtained empirical results confirmed that the invariance of the distribution of the number of accepted requests in the service system with respect to the service time distribution, which holds for classical DC-like systems with a Poisson incoming flow, does not persist in the case of a *BMAP*-flow. Thus, when estimating the probability of request loss in a real service system, the assumption that the incoming flow is a stationary Poisson process and that the service time follows an exponential distribution requires empirical validation.

Despite significant progress in the study of multi-channel DC-like systems with repeated calls, the problem of determining the stationary probability distribution of such models remains only partially resolved, even for classical *M*I*M*I*N*–type service systems. The first attempt to formalize the ergodicity condition for an *BMAP|PH|N*–type service system with repeated calls was partially addressed in [234]. However, the core issue (determining the stationary probability distribution of the service system states) remained beyond the scope of that study. Further development of this topic took place in [235], where a redialing strategy with an arbitrary functional dependence of the repeated request intensity on the number of requests in orbit was introduced into the *BMAP|M|N*–type model with repeated calls. Special attention was given to cases where the total intensity of repeated requests remains constant or grows indefinitely as the number of requests in orbit increases. The dynamics of the *BMAP|M|N*–type model were examined in [235] based on its asymptotic approximation to a quasi-Toeplitz Markov chain. This adaptation made it possible to determine the ergodicity conditions, stationary probability distribution, and key performance indicators of the *BMAP|M|N*–type service system. Subsequently, in [233], these results were generalized for more complex DC-like systems, where the service time follows a phase-type distribution. The study examined a multidimensional continuous-time Markov process, incorporating parameters such as the number of requests in orbit, the number of occupied service stations, and the state of the control processes of both the incoming and service flows for each occupied service station. The model employed dynamic channel numbering, which involved renumbering upon each service completion or initiation.

The mathematical model presented in [233] found practical application in [236] in the context of evaluating the performance of hot spots in airport wireless communication networks. In [237], a further generalization of the model from [233] was proposed for scenarios where orbiting requests are not entirely persistent. Specifically, the study accounted for the possibility that, after an unsuccessful attempt to gain access, some requests may, with a certain probability, cease further attempts and permanently leave the orbit without receiving service. The study [237] also included extensive numerical experiments, analysing the impact of incoming flow intensity, traffic correlation structure, and variations in inter-arrival times on the performance characteristics of the service system. The model presented in [233] was later utilized in the study [238]. The primary motivation behind this research was that the approach to numbering occupied service stations proposed in [233] led to a significant increase in the dimensionality of the block-generating matrices of the Markov process. In the case where the number of occupied channels is k and the state space of the service control processes has a dimensionality of Z, the total number of possible states equals Z^k for $k = \overline{0, N}$, leading to an exponential increase in the size of the block-generating matrices. For instance, when $S = 1, Z = 2, N = 5$, the size of the block-generating matrices is $(S + 1)\frac{Z^{N+1}-1}{Z-1} = 126$, whereas for $N = 7$, it already reaches 510. Given that, in a real information and communication system, the number of service stations is significantly larger, computing the characteristics of the corresponding model becomes challenging due to the limited availability of operational memory required for storing large matrices. To mitigate this constraint, an alternative Markov model was proposed in [238], where, instead of tracking the state of the service control process for each service station, only the number of service stations in the respective service phases was monitored. Within this assumption, [238] derived the stationary characteristics of the *BMAP|PH|N*–type model with repeated calls, and numerical experiments confirmed a reduction in computational costs compared to the traditional approach.

An important study is [239], where the model proposed in [233] was generalized to account for an unreliable service system. In this model, in addition to standard service processes, an *MAP*-flow of service station failures was introduced. The recovery period of service stations after failure was modelled using a phase-type distribution to incorporate complex stochastic repair mechanisms. In [240], a real call centre service system was represented by an *MAP|M|N*–type model with repeated calls. A key feature of this model was the presence of a finite buffer, providing an additional mechanism for temporarily storing requests before servicing. The study [241] also examined an *MAP|M|N*–type model with repeated calls, but in the context of servicing accepted requests in groups of a predetermined size. This model was designed to explore batch processing mechanisms for request handling, with a focus on collective access systems and resource management.

The studies [239–242] analysed a DC-like *MAP|PH|N*–type system with repeated calls, incorporating several specific characteristics. In particular, the model accounted for Markovian modulation of individual redialing intensities, allowing for the consideration of subscriber behaviour variability over time. Additionally, the model included a class of requests that may abandon further attempts to obtain service if their time spent in orbit exceeds a certain threshold. Finally, a service discipline was considered in which requests, when immediate service is unavailable, may decide to leave the system—an approach that effectively models the behaviour of real subscribers in high-load communication systems.

2.6 Custom DC-Like Systems with *BMAP*-Flow

The previous sections of this chapter examined the results of modelling information and communication systems within the framework of classical service system types, generalized for the case of *BMAP*-flow. This section focuses on complex, non-standard models, particularly tandem service system models, unreliable service system models (including those with catastrophic failures) DC-like system models with idle periods, and models of controlled service systems.

A tandem service system is characterized by multiple sequential service phases. In such a system, upon completing service at a given phase, a request transitions to the next service phase. These models serve as fundamental frameworks for describing the operation of service networks and are widely used in practice, particularly for modelling multi-step technological and communication processes.

The most extensively studied are dual tandem service systems, i.e. models consisting of two sequential service phases. The analytical analysis of such DC-like systems reduces to the study of a multidimensional stochastic process. In the simplest cases (rarely encountered in practice), the stationary state probability distribution of the tandem process exhibits a multiplicative structure, significantly simplifying its computation. However, when a closed-form analytical solution does not exist and the buffers at both tandem phases are unlimited, while the system operates under the assumption that the incoming flow is a stationary Poisson process and the service times at tandem phases follow an exponential distribution, determining the stationary state probability distribution reduces to a complex analysis of stochastic wandering in the quarter-plane. Analytically, this is expressed in the form of a functional equation for the two-dimensional generating function of the distribution. If at least one of the two buffers in a dual tandem system is finite, the incoming flow is an *MAP*-flow, and the service times follow a phase-type distribution, then the process describing the state of the tandem can be classified as a vector birth-and-death process. It was the practical interest in analysing dual tandem DC-like systems that motivated M. Neuts to develop the theory of vector birth-and-death processes [155].

The study [243] examined a tandem service system consisting of two queues with unlimited buffers, where the incoming request flow is an *MAP*-flow and the service times follow an exponential distribution. The dynamics of this model were characterized by a vector birth-and-death process, whose generator was represented as a matrix structure with a block architecture, where the block size grows indefinitely. In this model, the first component of the vector birth-and-death process determines the total number of requests across the two tandem phases, while the second component describes the possible distributions of these requests between the phases. Since the theory of vector birth-and-death processes is primarily developed for cases with fixed block sizes, the results obtained in [243] are largely heuristic. Greater practical applicability is found in dual tandem systems where at least one buffer is finite, the incoming request flow is an *MAP*-flow, and the service times are characterized by the following structure:

- At one of the phases, which has a single service station, the service time distribution is arbitrary, meaning it is not necessarily phase type.
- At the other phase, the service times follow a phase-type distribution or its special case—the exponential distribution.

Let us recall a distinct class of relatively simple dual tandem models for analysis, where the incoming flow is modelled as an *MAP*-flow (but not an *BMAP*-flow), and the restriction on the service time distribution applies only to the second phase, while the first phase has a finite buffer. The study of such models has been conducted in a series of works by A. Gómez-Corral [244]. A potential simplification of the analytical procedure for this class of models is based on the fact that the first phase of such a tandem can be interpreted as a classical *MAP|PH|1|N*–type service system. In this service system, the outgoing flow of requests arriving at the second phase is also an *MAP*-flow, and its matrix representation can be easily derived using matrices that describe the characteristics of the incoming flow and the service process.

If an arbitrary service time distribution is allowed at the first phase of a dual tandem, the analysis of the model becomes significantly more complex. When constructing an embedded Markov chain based on service completion moments at the first phase, its dynamics become substantially more intricate compared to the behaviour of a classical $M|G|1$–type service system. In such cases, a decomposition approach is typically applied to separately analyse the outgoing flow from the first phase of the tandem, which does not follow a *BMAP*-flow [245].

The exact analytical analysis of tandem systems, where the first phase is interpreted as a DC-like $BMAP|G|1$–or $BMAP|G|1$–type system, is conducted using the method of embedded Markov chains combined with the framework of Markov renewal processes. The study [246] examined transformations of $BMAP|G|1 \rightarrow PH|1|M-1$–and $BMAP|G|1|N \rightarrow PH|1|M-1$–type models, in which requests completing service at the first phase are lost if the buffer of the second phase is full. The study [201] examined a similar service system but under the condition that a request is not lost when the second phase buffer is full. Instead, it blocks the service station at the first phase until a slot becomes available in the second phase. In [247], a tandem model with an *MAP*-flow, finite buffers, and an arbitrary service time distribution at both phases was analysed. In this model, if the buffer of the second phase is full, requests that complete service at the first phase are lost. The study [248] investigated a similar service system but with the blocking of the first service station instead of request losses. In both cases, the analysis was conducted using the method of embedded Markov chains combined with Markov renewal processes, where the embedded process was focused on service completion moments at the first phase.

Since in most studies the service time distribution at the second phase is arbitrary, to ensure the Markov property of the embedded process, it is assumed that upon completion of service at the first phase, the service process at the second phase restarts. The study [238] examined a model similar to [236] but incorporating a feedback mechanism that allows requests that have completed the second phase to return to the first phase for reprocessing. In [249], a tandem variant similar to the one described in [236] was analysed, but without a buffer at the first phase. In this model, requests arriving when the service station at the first phase is occupied make repeated attempts to receive service. The study [250] analyses a tandem system where the first phase is described by a $BMAP|G|1$–type model, while the second phase is a multichannel, bufferless service system with exponentially distributed service times. A key feature of this model is that each request, after being serviced at the first phase, may require simultaneous service at multiple service stations in the second phase. If, at the moment of arrival, there are not enough available service stations, the request is either lost or temporarily blocks the operation of the service station at the first phase until the required number of resources becomes available. The study [251] provides a detailed analysis of this service system, including the computation of the distributions of request residence times at each phase separately and in the tandem system as a whole. The results are presented in a compact and numerically efficient form. The study [252] focuses on the analysis of a tandem system where, in addition to the main request flow, the second phase also receives handover traffic, which is serviced under the condition that a portion of the service stations is reserved specifically for this flow. The study [253] examines a tandem system where the first phase is modelled as a DC-like $BMAP|G|1$—type system, while the second phase is described by an arbitrary Markov process with a finite state space. This approach allows for the modelling of complex multi-channel service systems with stochastically controlled service parameters.

There are very few precedents for studying tandem service systems with *BMAP*-flow that include more than two phases. In [254], an approximate method was proposed for analysing a tandem system composed of a finite number of single-channel service systems, where service times follow a phase-type distribution and incoming requests are generalized as a *BMAP*-flow. The impossibility of an exact analysis of such models is explained by the fact that, in the presence of unlimited buffers, the output flows from each service system within the tandem lose the structure of an *MAP*-flow. Although these output flows formally approximate *MAP*-flows, their control processes have an infinite state space, making it impossible to apply the classical mathematical framework of vector birth-and-death processes to their respective service systems. This is because only a limited set of operations can be performed with infinite-dimensional matrices.

In [255], an analytical approach was proposed for studying a tandem system composed of a finite number of multi-channel DC-like systems without buffers, where service times follow an exponential distribution and the request flow is characterized as *MAP*-like. The latter assumption involved treating the output flow from each phase as *MAP*-flows, followed by determining the stationary probability distribution of a generalized $MAP|M|N|0$–type model. This approach made it possible to obtain the joint state distribution both for selected phase sequences within the tandem and for the tandem as a whole, as well as to derive analytical expressions for the probabilities of request loss in the service system. The methodology formulated in [255] was generalized in [256] to tandems where each phase additionally receives an external *MAP*-flow. A similar approach was applied in [151, 163] to study a tandem composed of a finite number of single-channel service systems with phase-type service time distributions, finite buffers, and external traffic mixed with the primary *MAP*-flow of requests. This model effectively describes processes in real broadband wireless networks, which generalize the sequence of heterogeneous transmission segments between subscribers and end service stations. Specifically, the model in [163] is focused on the analysis of broadband wireless sensor networks. In the context of this research object, the study describes the process of information transmission from numerous sensors to a central control unit, incorporating cybersecurity elements through video surveillance of selected critical infrastructure components and supporting voice IP communication between mobile and stationary communication points.

The primary obstacle to the exact analysis of multiphase tandem service systems is the exponential growth of the state-space dimension of the generalized multidimensional continuous-time Markov process as the number of tandem phases increases. As a result, for a large number of phases, direct analytical computation of service system characteristics becomes challenging. To address this issue, an approximation method has been proposed, wherein the *MAP*-flow exiting each tandem phase is approximated by an *MAP*-flow of lower dimensionality while preserving its key statistical properties, including intensity, correlation, and variation in inter-arrival times. This approach allows for limiting the state-space dimensionality while ensuring an acceptable level of computational accuracy. Another effective method is the transition to simulation-based models, which enable the analysis of complex tandem DC-like systems with a large number of phases. Specifically, in [163], a tandem service system with 20 service phases was proposed and analysed using simulation modelling, demonstrating the effectiveness of this method for studying highly structured DC-like systems of high complexity.

The analysis of unreliable service systems with *BMAP*-flow is a crucial component of queueing theory, as it reflects the common real-world scenario in which service stations periodically fail and require repair or replacement. This characteristic is inherent in all real technical and communication systems, which explains the significant research interest in studying such models. Let us focus on a class of unreliable service systems where both the request flow and the failure flow of service stations are modelled as *MAP*-flows. [257] is one of the earliest studies dedicated to the analysis of unreliable $BMAP|G|1$—type service systems with repeated calls using the matrix-analytic method and matrix factorization. Further development of this topic is presented in [223], where a multi-channel DC-like $BMAP|PH|N$—type system with repeated calls was analysed. This model is characterized not only by the fact that the request flow is described by an *BMAP*-flow but also by the fact that the failure flow of service stations follows an *MAP*-flow, while the recovery time of service stations after failure follows a phase-type distribution. A distinctive feature of this model is that requests arriving when all service stations are occupied do not permanently leave the DC-like system but instead make repeated attempts to obtain service. The results obtained in [223] can be easily generalized to service systems with an infinite buffer by formally considering the limiting case of queue length growth.

In the context of analysing information and communication systems of critical use, a particularly interesting class of unreliable service systems is those susceptible to induced failures or “attacks”. In such models, service station failures result from the arrival of a special “attacking” group *MAP*-flow, which causes the failure of one of the occupied service stations and the loss of the request being serviced. Each attack also triggers a mechanism for subsequent failures of various types, which are modelled as individual *MAP*-flows. The recovery of a service station in such DC-like systems depends on the type of failure and is described by a phase-type distribution with parameters specific to each failure class.

In [258], the problem was formulated and solved not only for determining traditional performance metrics of service systems with failures (such as average queue length and average waiting time) but also for an important dependability indicator—survivability. This indicator characterizes the ability of a service system to continue functioning effectively under systematic failures, including induced ones. This significantly expands the practical applicability of the theoretical results from [258] to the performance analysis of real-world complex technological and communication systems.

A particularly interesting aspect is the comparative analysis of the class of unreliable DC-like systems and the class of service systems with induced failures caused by attacking requests. Models belonging to these classes share significant similarities in the sense that the arrival of an attacking request is analogous to a sudden service station's failure, as in both cases, the servicing of the current request is interrupted. However, the fundamental difference lies in the fact that, in the case of a service station's failure, the service station transitions into a repair state and is temporarily unavailable for servicing subsequent requests. In contrast, when an attacking request arrives, the service station remains operational and continues servicing new requests.

In [259], an analysis was conducted on a multi-channel DC-like $BMAP|PH|N$—type system, where an MAP-flow of attacking requests operates in such a way that, upon arrival, accepted requests are removed from the service system through a specific eviction mechanism. This mechanism functions as follows: an attacking request entering the service system "annuls" an ordinary request from the end of the queue, if one exists. This means that, unlike in the classical failure scenario, the servicing of a request on a service station is not interrupted; instead, the waiting queue is reduced. An alternative approach to modelling DC-like systems with attacking requests was considered in [260], where a $BMAP|PH|N$—type model was analysed with an MAP-flow of attacking requests capable of annulling (interrupting) the active servicing of an ordinary request [260]. A distinctive feature of [260] is the consideration of scenarios where an attacking request may be ignored if, at that moment, a non-ordinary (protected) request is being serviced in a special state. In this case, the attacking request leaves the service system without affecting the protected request in service.

A significant contribution to the study of single-channel and multi-channel DC-like systems with attacking requests is also presented in [260, 261]. These studies analysed the impact of attacking requests on key service system performance metrics, such as the probability of ordinary request loss, the average time a request spends in the service system, and other standard characteristics. These metrics are crucial for the practical application of the obtained results.

A distinct subclass of attacking requests in unreliable service systems consists of "catastrophic" requests. Unlike ordinary attacking requests, which remove only a single request from the service system, a catastrophic request results in the instantaneous removal of all accepted requests, completely clearing the system. The arrival of a catastrophic request is modelled with the consideration that, during its processing, the system's service station may either fail or remain operational. Among the studies dedicated to the analysis of DC-like systems with a $BMAP$ flow of ordinary and catastrophic requests, notable contributions include [262, 263]. In [262], the vector-generating function method was applied to analyse a $BMAP|SM|1$—type model with an MAP-flow of catastrophic requests. The study [263] builds on the generalization of M. Neuts' approach, which enabled the development of more numerically stable algorithms for computing the stationary probability distribution of the corresponding service system compared to the vector-generating function method. In both [262] and [263], it is assumed that the arrival of a catastrophic request immediately removes all requests from the service system but does not cause the failure of service stations. In contrast, [264] examined a modified model where a catastrophic request leads to the failure of service stations, and the recovery time is interpreted as a stochastic variable with an arbitrary distribution. This study also considered scenarios where ordinary requests arriving during the repair process may either be lost or accumulate in a buffer until the service system is restored. A further generalization of approaches to the study of DC-like systems with catastrophic requests was carried out in [265]. This study proposed an analytical framework based on multidimensional Markov chains to analyse a multi-channel service system, where the incoming *BMAP*-flow of ordinary requests, the flow of attacking requests, and the flow of catastrophic requests are statistically correlated.

Within the broader class of unreliable DC-like systems, an even more general category is that of service systems operating in a variable environment. Their key characteristic is that the parameters governing system operation (including request arrival rates, service intensities, redial attempts, failure and recovery rates of service stations, and other factors) depend on an external Markov process, which describes the stochastic changes in external conditions affecting the service system. When the state of this process changes, the parameters of the DC-like system instantly adjust according to the current state of the external environment. The literature presents a wide range of service systems with *MAP*- and *BMAP*-flows operating in a dynamic environment. Significant findings on the analysis of such models are provided in several studies [267, 268]. A crucial feature of the latter among them is the consideration that, when the state of the external environment changes, not only the characteristics of input and output flows or service durations may be affected in the service system, but also buffer capacities, retry policies, and the intensity of failures and recoveries. Notably, studies [268, 269] explore adaptive DC-like systems, in which the number of service stations or buffer capacity dynamically adjust depending on the state of the external Markov process.

Service systems with unreliable service stations are significantly similar to DC-like systems, in which periodic rest periods for service stations (so-called "server vacation" systems) are incorporated. The common features of these models lie in the fact that, in both unreliable service systems and service systems with rest periods, service stations are temporarily unable to process requests. However, a fundamental distinction exists in the nature of the cause leading to the temporary unavailability of the service station. In unreliable systems, this occurs unexpectedly due to an unforeseen failure, whereas in models with rest periods, service station downtime is planned and intentional, adhering to information and functional security policies.

Service systems with service station rest periods hold significant practical value, as the ability to schedule breaks, for instance, for preventive maintenance, is a crucial element in the operation of most real-world technical systems. Examples of applications of such models with rest periods include networks with periodic technological shutdowns and information and communication systems with organised low-activity periods. Additionally, the study of models with rest periods is essential for analysing multiphase tandem systems with multiple buffers, where the service station sequentially processes requests from different queues, implementing the polling principle. The scientific literature provides a rather detailed classification of systems with service station rest periods [270]. This classification is based both on the strategies governing the departure of service stations for rest and on the conditions for their return to operation after the rest period. The most common and well-studied strategy involves the service station entering a rest period only after the service system has been completely cleared of requests.

The study of service systems with *BMAP*-flows and service station rest strategies can be effectively implemented using theoretical developments derived for the analysis of multidimensional quasi-Toeplitz Markov chains. In the context of modelling systems with rest periods, the structure of the block transition probability matrix of type (2.9) remains almost unchanged, with only the specific form of the blocks in the first row B_m, $m \geq 0$, requiring adaptation to the particular characteristics of the real system under investigation. Several studies have analysed service systems with *BMAP*-flows and service station rest periods. Among the early studies of such models, works [271, 272] stand out, providing a detailed examination of various scenarios and operational characteristics of service systems with rest periods, including cases of adaptive management of rest modes depending on workload or external conditions. A significant contribution to the development of this field comes from research on cyclic systems, where service station rest periods are employed to regulate the service process of different request flows competing for the resources of a single service station. Additional findings and a comprehensive review of the issues related to DC-like systems with service station rest periods are presented, in particular, in [270], which compiles both classical and recent results that have formed the foundation for further theoretical research and practical applications of such models.

Among the recent studies on service systems with service station rest periods, notable contributions include [273], which examines the case where the maximum number of consecutive rest periods is predetermined; [274], which considers a model incorporating the possibility of dynamically adjusting the service rate for requests; [172], which proposes a generalized strategy for organizing rest periods, taking into account an *MAP*-flow of attacking requests; and [275], which analyses a service system with a limited service discipline, where the service station enters a rest period not only after the system has been cleared of accepted requests but also upon reaching a predefined threshold of serviced requests. Another type of dependence of the rest period on the prior activity of the service system is explored in studies [273–276]. Specifically, study [276] focuses on the impact of the classical "exhaustive" service strategy on the functioning of a service system with rest periods, whereas studies [131, 132] examine a more complex "gated" service strategy, in which the service station, upon each activation, processes only the requests present in the buffer at the moment of activation. Study [277] examines a cyclic polling system consisting of multiple buffers, each modelled as a single-line service system of type $BMAP|G|1$. Recently, a significant number of publications have emerged in the scientific literature focusing on so-called "working vacation" service systems, which are characterized by the fact that during rest periods, service stations continue to process requests but at a reduced intensity.

This subsection concludes with a concise analysis of recent studies on controlled DC-like systems. A typical example of this class of service systems is a model in which certain stochastic parameters or probability distributions characterizing the operation of a real system can change depending on the current state of the model or other external factors. Such service systems are particularly characterized by the dynamic adjustment of probabilistic service process parameters, input flow intensities, or access strategies. Representative studies covering various aspects of the analysis of controlled DC-like systems include [278–280].

The most relevant topic within the scope of this section is the study of communication resource reservation in information and communication systems within the formalism of controlled DC-like models. One of the key aspects of cellular mobile network operation is ensuring service continuity for users during transitions between coverage areas (handover). Specifically, connection loss during cell switching is a critical issue with significantly more severe consequences than the inability to establish a new connection due to a lack of available channels. To reduce the probability of such disconnections, mechanisms for prioritized channel access must be implemented for subscribers already in service. One of the most effective mechanisms for ensuring priority to users undergoing handover is the channel protection strategy, also known as the Guard Channel Policy [281]. Its core idea is to reserve a portion of the channels exclusively for servicing subscribers transitioning to a new cell. If the total number of channels in a cell is N, then $R < N$ channels are allocated specifically for users performing a handover (hereafter referred to as first-type users). New users (second-type users) can access the network only if, at the moment of their request, the number of occupied channels does not exceed $N - R$. If all available channels are occupied, a new request is either lost or placed in a waiting queue (orbit) and attempts reconnection at a later time.

The efficiency of various approaches to resource reservation in information and communication systems has been analysed in numerous studies. For instance, [282] examines a specific case where only a single channel is reserved, i.e. $R = 1$. The numerical experiment conducted in this study confirmed that even such a minimal reservation significantly improves the service quality for first-type users. In the general case, the problem of optimal selection requires an analysis of the trade-off between the loss probabilities of both types of requests and the channel utilization level. A brief overview of recent research on this issue is provided in [123]. Study [283] stands out among similar works as it offers a mathematically precise analysis of a model that accounts for different service rates for first- and second-type users. However, a limitation of this model is its disregard for the effect of repeated call attempts and its assumption of a stationary Poisson process as the model for the incoming request flow. Study [284] accounts for the effect of repeated call attempts and employs the *MAP*-model to characterize the incoming flow. However, this work does not consider channel reservation, and the maximum number of subscribers in the cell remains limited by a fixed constant. Study [238] is based on a more general model that incorporates a phase-type distribution for both service duration and inter-retry intervals. However, similar to [237], this model imposes a restriction on the maximum number of subscribers in the cell. Study [26] analyses a system similar to [238] but introduces an additional assumption that the total intensity of repeated attempts is independent of the current number of requests in the orbit, which does not accurately reflect the realities of information and communication networks. Study [239] focuses on the analysis of a DC-like system similar to the one examined in [123], but with two critical constraints: the service system is based on Poisson flows and assumes identical service duration distributions for both types of requests. Similar simplifications are employed in [125], where, as in [126], it is assumed that the intensity of repeated attempts is independent of the number of requests in the orbit. All the aforementioned limitations have been addressed in the model presented in [123]. This model accounts for *MAP*-flows for both types of users, as well as distinct service duration distributions for requests of different classes. Additionally, the intensity of repeated attempts is modelled proportionally to the number of requests in the orbit, providing a more accurate representation of real-world systems. In this study, the conditions for ergodicity and the stationary state probability distribution of the service system are formalized using the framework of asymptotic quasi-Toeplitz Markov chains. Numerical analysis methods have demonstrated the high efficiency of the channel reservation strategy under an optimal selection of the parameter R. In study [127], the model presented in [123] was significantly improved by incorporating several realistic assumptions that provide a more accurate representation of the functioning of DC-like systems with controlled access and repeated attempts. Specifically, the following generalizations were introduced into the baseline model:

- The model incorporates an additional intermediate buffer of limited capacity, designated for subscribers undergoing handover. This buffer temporarily stores requests from mobile users transitioning between adjacent cells with partially overlapping coverage areas in cases where no free channels are available in the new cell.
- The assumption of subscriber impatience in the intermediate buffer is introduced. This means that a mobile user can independently terminate waiting in the intermediate buffer and leave the current cell, either moving to another cell or ending the connection.
- It is assumed that a subscriber whose request is stored in the intermediate buffer may be transferred to the orbit of repeated calls if a channel does not become available. This extends the classical retry mechanism, allowing for hybrid scenarios in the operation of the information and communication system.
- Requests in the orbit may exhibit impatience and lack persistence, meaning that after several unsuccessful retry attempts, they may leave the orbit without receiving service, with a certain probability.
- The service duration distribution in the proposed model is more general than the classical exponential law and is characterized by a phase-type distribution, enabling a more accurate representation of real information and communication system operations.

Finally, a promising class of session-based arrival models for controlled DC-like systems should be mentioned, where requests do not arrive individually but in the form of sessions, with parameters such as arrival intensity and session duration depending on the current state of the service system or external factors. The analysis and optimization of such systems have been explored in studies [240–247], yet many questions remain unresolved.

2.7 Conclusion

The modelling of information and communication systems of critical use is a pressing issue due to the necessity of ensuring stability, efficient resource management, and minimizing losses in networks operating under high-load conditions. This study examines DC-like systems, which facilitate the description of service processes while accounting for the stochastic nature of requests, batch arrivals, dynamic resource reservation, and adaptive data flow management.

Section 2.2 presents the mathematical framework of BMAP flows as a generalization of classical request arrival models in service systems. BMAP flows enable the consideration of batch request arrivals and their interdependence, which is crucial for analysing communication networks where traffic loads exhibit a clustered structure. The proposed model allows not only for the evaluation of fundamental traffic characteristics but also for the derivation of analytical expressions for key performance indicators.

Section 2.3 explores two-dimensional Markov chains as a foundation for modelling dynamic processes in service systems with BMAP flows. The proposed approach formalizes the reception and servicing of requests in multi-channel networks, enabling precise evaluation of system state probabilities and forecasting of load dynamics. The application of matrix-analytic methods ensures high accuracy in performance estimation for systems with varying degrees of correlation between incoming request flows.

Section 2.4 provides a detailed analysis of single-channel DC-like systems with BMAP flows, considering variations of models with both unlimited and limited buffer space. The impact of service time distribution on request loss probability, average waiting time, and system load is investigated. It has been established that employing phase-type service time distributions enhances model adaptability to real-world network conditions.

Section 2.5 analyses multi-channel systems with BMAP flows, which facilitate the modelling of service processes in networks with dynamic resource allocation. Service management strategies are examined for scenarios where resources are constrained, and batch arrivals occur. Three approaches are proposed: partial admission, complete admission, and complete rejection, each offering distinct advantages depending on the target system performance. Numerical experiments demonstrate that adaptive buffering management, which accounts for current system load and request priority, yields optimal results.

Section 2.6 investigates adaptive queue management mechanisms that enhance system performance under variable loads. The study examines methods for dynamic buffer resizing, priority redistribution, and queue management algorithm integration based on real-time network conditions. The proposed approaches significantly improve system efficiency in cases of high request variability.

The generalization of the obtained results confirms the effectiveness of extended stochastic models in managing resources within information and communication systems of critical use. It has been found that the integration of BMAP models with

Markovian methods enables precise load forecasting, optimal network parameter determination, and minimization of request loss probability.

Future research directions include the refinement of adaptive buffer and queue management mechanisms to optimize resource allocation under variable loads. A key aspect is the development of dynamic buffers with variable sizes that adapt to real-time system load, preventing overloads. Context-aware queue management, based on historical data analysis and load forecasting, will contribute to improved request handling. Additionally, the advancement of machine learning-based priority servicing policies will ensure optimal resource allocation between critical and non-critical requests. Further investigation into intelligent request dispatching algorithms will aid in minimizing average waiting times through efficient load balancing.

A significant challenge lies in modelling multi-channel systems with heterogeneous traffic flows, as real-world information networks service requests with diverse characteristics. It is necessary to develop universal multi-channel service models that consider traffic variability and ensure even load distribution. Research into non-homogeneous BMAP flows will provide insights into how external conditions affect request arrival intensity. One promising direction is the development of flexible channel allocation strategies based on current load levels, which will minimize delays and enhance system throughput. Furthermore, the optimization of asynchronous flow servicing in distributed information and communication networks requires the extension of classical mathematical models.

The integration of machine learning methods into resource management systems opens new opportunities for improving network adaptability and performance. The use of neural network-based load forecasting algorithms will enable dynamic parameter adjustments in response to traffic fluctuations. Reinforcement learning techniques can be applied for automatic load balancing in high-load networks, significantly enhancing the efficiency of resource utilization. AI-based data stream analytics will facilitate anomaly detection and overload prevention in real-time systems. The integration of deep learning into dynamic resource scaling mechanisms in cloud and mobile networks will ensure adaptability to environmental changes and enhance operational stability.

A promising research direction involves the application of stochastic geometry for analysing next-generation network (6G) performance, allowing for the incorporation of spatial factors and their impact on resource distribution. The development of stochastic models for spatial network node placement will enable more accurate system performance predictions. The analysis of spatial correlation in traffic patterns will help anticipate local overloads and improve load distribution across networks. The application of random graph theory will provide insights into the reliability and scalability of distributed networks, which is critical for 6G systems. Additionally, the modelling of holographic communications and intelligent reflecting surfaces (IRS) is expected to play a key role in future wireless technologies.

Finally, a crucial area of research is the development of energy-efficient strategies for information and communication systems, which will reduce energy consumption without compromising performance. The study of energy-efficient queue management algorithms will help decrease infrastructure load while maintaining service quality. Optimizing data transmission in wireless networks through dynamic power regulation of transmitters will minimize energy losses and enhance system stability. The implementation of Green ICT technologies, particularly the integration of renewable energy sources for powering network nodes, will improve the ecological efficiency of information systems. Lastly, predictive energy consumption control methods will ensure optimal resource utilization in cloud and edge computing, shaping the future of telecommunication networks.

Chapter 3
Simulation Study of Divide-and-Conquer Models Under 5G Infrastructure Conditions

3.1 Introduction to the Simulation Study

In modern 5G-IoT infrastructures, where data processing is parallel, asynchronous, and highly dynamic, analytical assessments of information system performance increasingly encounter the limits of their applicability. The divide-and-conquer (DC-like) formal models forming the foundation of this monograph are developed to overcome the constraints of classical service queueing systems by introducing flow coordination, request splitting, and result synchronisation. However, even the most accurate analytical tools require validation through practical scenarios, particularly in cases involving correlated input flows, stochastic failures, scaling in multi-server and cascade networks, as well as the impact of updates and mechanisms of limited fault tolerance.

The foundation for the simulation study in this Appendix is based on the models formulated in detail in Sects. 1.2–1.4 of the main text of the monograph. Each of these represents a distinct class of DC-like systems within 5G-IoT environments, characterised by specific requirements for fault tolerance, scalability, and load management. The model presented in Sect. 1.2 simulates a system with two classes of incoming requests (regular and attacking), with the latter granted absolute priority and incorporating a stochastic update mechanism. This configuration emulates information systems operating under cyber threats with reactive queue clearing [31]. Formulae (1.45)–(1.47), derived using the generating function method and the matrix-geometric approach, define the mathematical structure of this model. Section 1.3 describes a cascade connection of two Erlang-type queues with finite service device capacities and individual generalised update mechanisms. This architecture enables the modelling of real-world scenarios involving two-factor data processing or authentication. The verification of analytical formula (1.35), which is based on level-phase analysis in a stochastic environment using intensity matrices, constitutes one of the key objectives of this study. The model from Sect. 1.4 focuses on cloud-based DC systems where functional availability and mean time to recovery

V. Kovtun, *Formal Models of Divide-and-Conquer Queuing Systems in 5G Infrastructures*, https://doi.org/10.1007/978-3-032-06187-4_3

are critical. In this case, the system is described in terms of high-level fault tolerance, and the simulations aim to confirm the accuracy of formula (1.21), which draws upon the concept of order statistics and combined availability probabilities. Each model is grounded in its own mathematical foundation: generating functions, matrix-geometric approach, order statistics theory, or extreme value estimation methods [294]. This allows not only for the verification of numerical results but also for the evaluation of the practical effectiveness of the corresponding approaches in modelling tasks. The correspondence between the analytical models and the simulation subsections is as follows:

- the model with a cyberattack (Sect. 1.2) → Sect. 3.3;
- the cascade network (Sect. 1.3) → Sect. 3.4;
- the cloud system (Sect. 1.4) → Sect. 3.5.

The simulation study presented in this Appendix aims not only to confirm the adequacy of the mathematical model formulations but also to investigate the specific features of their behaviour under varying conditions of load, request correlation, changes in service time, and architectural complexity. Each experiment is designed to identify the factors that significantly influence key performance characteristics of the system, including response speed, resilience to overload, occurrence of losses, and level of availability. Particular attention is paid to the sensitivity of systems to the type of input flow (Poisson or BMAP), changes in the service time distribution (exponential, Erlang, hyperexponential, Cox), as well as to the scaling of the number of servers, synchronisation of results, and behaviour under update or queue-clearing mechanisms. The experiments make it possible to trace both average and extreme effects, including scenarios that drive the system towards an unstable regime or functional collapse.

The selected metrics encompass both temporal and probabilistic characteristics of the system, enabling a comprehensive assessment of its performance, stability, and quality of service under critical conditions. The following set of control metrics is defined:

- Average Response Time $E[T]$: the mathematical expectation of the total time a request spends in the system, from its arrival until the completion of processing of all its sub-requests;
- Loss Probability p_{loss}: the proportion of requests that were not processed due to buffer limitations or queue clearing during updates;
- Queue Instability Function $\Delta Q(t)$: quantitatively evaluates the temporal variability of the queue length, measuring the degree of fluctuation in the buffer size at time t;
- Tail Latency (95th Percentile) T_p: the delay exceeded by only 5% of requests; a key indicator in systems with QoS guarantees, particularly for real-time traffic;
- Effective Queue Flushing Intensity ν_{ef}: the average number of queue flush operations (complete or partial) per unit time under stochastic or conditional update mechanisms;

- Average Synchronisation Time $E[S_{\text{sync}}]$: the expected duration for which processed sub-requests remain in the buffer before being merged.

All metrics are consistent with the theoretical characteristics analysed in the corresponding sections of the monograph. Their calculation is carried out both on the basis of simulation data and through analytical forecasting using formulae (1.13), (1.21), (1.35), (1.46), which will be the subject of a dedicated comparison in Sect. 3.6. It is anticipated that different configurations of flows and service time distributions will affect the mentioned metrics in various ways: for instance, a BMAP flow with high burstiness may increase tail latency and losses, whereas the Cox distribution may reduce the average synchronisation time due to phase concentration. Such analysis enables the identification of model sensitivity to specific operational conditions and facilitates the formulation of recommendations for their practical application.

Evaluating the behaviour of DC-like systems in realistic scenarios requires careful selection of input flow and service time models that enable control over randomness, correlation, and variability. In this simulation study, particular attention is devoted to comparing Poisson and BMAP flows, as well as analysing service models with exponential, Erlang, hyperexponential, and Cox distributions. The Poisson flow, being Markovian and memoryless, serves as a benchmark model describing random events with independent intervals. Its intensity is denoted as λ, and it forms the basis of classical M/M/1-type systems. At the same time, in real-time systems, traffic is often bursty, packet-based, or governed by external activation patterns—characteristics that BMAP is designed to model. This flow allows not only the specification of the average arrival intensity λ, but also its variance, autocorrelation, and packet structure, making it well-suited for modelling 5G-IoT loads. All experiments in this Appendix are conducted at the same average intensity λ for both types of flows, thereby isolating the influence of structural differences rather than variations in load.

As for the service models, four representative types of distributions were selected:

- The exponential distribution as the baseline case with the memoryless property, which provides the simplest analytical framework, particularly for M/M/1-type models.
- The second-order Erlang distribution as a model with a fixed number of service phases, which reduces variability compared to the exponential case, thus enhancing processing predictability.
- The hyperexponential distribution, on the contrary, allows for modelling overvariability with long queues and significant delays for individual requests, which is critical for assessing T_p.
- The Cox distribution generalises the Erlang structure by enabling random transitions between phases, offering flexible control over the shape of the service time density and allowing the modelling of both narrow-profile and extended scenarios. Due to its parametric flexibility, this distribution can approximate a wide class of phase-type distributions, including non-Poisson scenarios with asymmetric phases and a random number of stages.

These distribution types were selected due to their capacity to capture the principal behavioural variants: from the simplest randomness to complex phase structures and highly variable delays. Other distributions, including deterministic or heavy-tailed ones, will be considered in future research. It should be emphasised that in the context of DC architectures with result synchronisation, both the type of input flow and the form of service time significantly affect interference between parallel queues, which in turn impacts the values of T_p, p_{loss}, and $E\left[S_{\text{sync}}\right]$. The selection of combinations of flow and service types in each simulation scenario is guided by the specific objectives of Sects. 3.2–3.5, while the results are systematised and summarised in Sect. 3.6.

The simulation study is conducted using discrete-event simulation (DES), which enables precise reproduction of the temporal dynamics of requests, state transitions, delays, and updates. The models are implemented in parallel within MATLAB R2023b and Python 3.11 (SimPy 4.0.2) environments to ensure platform-independent results and to enhance visualisation flexibility.

In MATLAB:

The following toolboxes are used:

- SimEvents (R2023b) for constructing block-based models with event-driven control;
- Statistics and Machine Learning Toolbox for generating second-order Erlang, hyperexponential, and Cox distributions;
- Symbolic Math Toolbox for analytical computation of control formulae;
- Parallel Computing Toolbox for parallel execution of ≥5000 simulations.

Key functions include

- *randg*, *exprnd*, *random*(Hyperexp,...), *makedist*(PhaseType,...) for generating random service times;
- *sim*, *set_param*, *get_simulink_block_handle* for running event-driven models in the SimEvents environment;
- *timetable*, *findpeaks*, *movmean* for analysing stabilisation during the warm-up period;
- *parfor* for accelerating simulations when examining scaling effects.

In Python:

The main framework used is SimPy 4.0.2, with additional modules:

- *numpy.random.Generator* for generating Poisson and BMAP flows using a custom implementation of matrices D_0, D_1;
- *scipy.stats* for modelling service times: *expon*, *erlang*, *hypergeom*, and custom phase-type distributions (Cox);
- *matplotlib.pyplot*, *seaborn* for plotting;
- *pandas.DataFrame* for storing and filtering event logs, constructing empirical distributions and computing T_p;
- *concurrent.futures.ProcessPoolExecutor* for organising parallel simulation runs.

The BMAP flow is implemented via an event generator with a Markovian transition structure defined by the matrix pair (D_0, D_1) [36], with synchronous batch emission of requests according to a probabilistic batch size vector.

General settings:

- Number of Repetitions per Scenario: $N = 5000$;
- Warm-up Conditions: stabilisation of the moving average of response time over an interval of length 100.

Within each scenario, the following parameters are experimentally varied: input flow intensity λ, average service time $1/\mu$, number of servers m, type of service time distribution, as well as the parameters of update and queue-clearing mechanisms. In cases where the models feature cascade architecture, multi-class traffic, or update mechanisms, extended queue components are implemented with priorities, clearing, controlled loss, and synchronisation buffers. These components correspond to the formal structures described in Sects. 1.2–1.4 of the monograph.

One of the key objectives of the simulation study is the verification of the analytical dependencies formulated in Sects. 1.2–1.4 of the main text of the monograph. In particular, each of Sects. 3.3–3.5 examines the accuracy of calculations for core metric characteristics by comparing them with the results of discrete-event simulation. All comparisons are performed under the condition of equivalence in input parameters, enabling the alignment of numerical values derived from analytical expressions with empirical estimates.

The deviation of empirical results from theoretical estimates is assessed both by absolute error and by relative bias. For each metric, confidence intervals with a 95% confidence level are additionally constructed, ensuring the statistical reliability of the observed effects. The comparison results are presented in the form of diagrams and plots, enabling conclusions to be drawn regarding the accuracy and applicability limits of each formula, as well as identifying scenarios in which analytical models require refinement or supplementation through numerical methods.

3.2 Reference Models: SM and ISM Systems

Two baseline models of divide-and-conquer systems have been selected as reference configurations for the simulation study. These models implement the splitting of an incoming request into parallel processing branches followed by result synchronisation: SM (Split-Merge) and ISM (Independent Split-Merge). Both models describe architectures without updates, failures, or aggressive interruptions, and are used as benchmark systems for subsequent comparisons with more complex variants in Sects. 3.3–3.5.

The formal characteristics of the SM and ISM models are described in Sects. 1.1.1 and 1.1.2 of the main part of the monograph. These sections present their structural organisation, typical servicing algorithms, and schemes for synchronisation time distribution. In this subsection, these models are employed as practical configurations for constructing a simulation benchmark. In the SM model, each request arriving at the system is split into k sub-requests, which are processed in parallel across k independent queues. The system is considered ready to respond only when all k sub-requests have been completed. As a result, the moment of service completion is determined by the maximum of the completion times of all sub-requests, leading to the "delay until the last" effect. In contrast, the ISM model implements independent fixed synchronisation. After sub-request processing, the results are placed into a buffer, where they are held for a predefined time $\theta\theta$, after which they are automatically merged into a response. This allows for the modelling of situations in which responses are generated even without the completion of all branches, simulating real-world constraints on processing time or tolerance for partial data loss. In Sect. 3.2, the parameter θ is fixed, but in subsequent analyses in 3.4, it may be varied to assess the system's sensitivity to synchronisation timeout duration.

The model parameters are defined as follows: intensity of the incoming Poisson flow λ (single-class); service rate at each server μ; number of parallel channels k (in simulations: $k = 3$, as this number of parallel channels corresponds to typical architectures of aggregated processing services in small-scale edge or cloud environments); fixed synchronisation interval in ISM θ (typical value: $\theta = 0.3$); actual delay S_{sync} between the completion of service and the formation of the response.

This subsection examines the baseline case: the request flow is a Poisson process with intensity $\lambda \in [0.1, 0.9]$, and the service time follows an exponential distribution with $\mu = 1$. All other parameters are fixed according to the simulation methodology described in 3.1. Both models are implemented in parallel within the MATLAB (SimEvents) and Python (SimPy) environments. In MATLAB, queue elements with a configured number of servers and parallel event simulation with synchronisation delay are used; in SimPy, parallel processes wait in a buffer until the merging event occurs. For ISM, a forced service completion mechanism is implemented via a timer to ensure a constant value $S_{\text{sync}} = \theta$.

Simulation modelling is conducted for the reference models SM and ISM under Poisson load with intensity $\lambda \in [0.1, 0.9]$ and exponential service time $\mu = 1$. Both models use $k = 3$ parallel channels, while ISM employs a fixed synchronisation timer $\theta = 0.3$. Each metric value is averaged over 5000 simulation runs with system warm-up. For the percentile T_p, the empirical 95th quantile of the response time distribution is constructed.

In the baseline load region $\lambda < 0.5$, both models exhibit similar values of $E[T]$, indicating the absence of delays caused by queue competition. In SM, the curve begins to accelerate after $\lambda > 0.6$, although the increase remains moderate and does not display exponential escalation: the growth results not from system degradation but from the gradual accumulation of collisions between sub-requests. In ISM, noticeable stabilisation is observed: following an initial increase at $\lambda < 0.4$, the average response time reaches a plateau that persists until the end of the interval. This plateau is due to the timer θ, which limits the impact of slow services. The behaviour of ISM reflects the expected stability under high-load conditions [36].

SM exhibits a gradual increase in T_p with a slight curvature after $\lambda = 0.6$, indicating a heightened probability of realisations with long waiting times due to the lag of a single branch. The percentile does not rise sharply, but sufficiently enough to negatively affect requests with critical time constraints. In ISM, the percentile increases almost linearly across the entire interval and never exceeds 1.5× the baseline level. The timer θ eliminates excessively slow realisations while preserving complete responses in typical cases. This demonstrates the advantage of ISM in maintaining stable QoS while adhering to computational complexity constraints.

In SM, the curve $E\left[S_{\text{sym}}\right]$ increases logarithmically, reflecting the gradual rise in the time required to merge results due to channel asynchrony. Even under high load, the increase does not become critical: average values remain within acceptable bounds without sharp jumps. For ISM, the values of $E\left[S_{\text{sym}}\right]$ remain consistently close to the level $\theta/2 \approx 0.15$, regardless of the load. This is a direct consequence of the deterministic timeout structure and confirms that the synchronisation buffer operates independently of the internal queue dynamics.

Overall, the simulations presented in Figs. 3.1, 3.2 and 3.3 confirm that both models remain operational under moderate load conditions; however, ISM exhibits greater predictability and stability when approaching boundary regimes. Its ability to constrain the tail of the response time distribution positions it as a reference architecture for comparison with adaptive systems in subsequent sections.

The results presented in Figs. 3.1, 3.2 and 3.3 provide a visual reference framework for the subsequent comparative analysis of models in Sects. 3.3–3.5. This subsection focuses on a fixed load value $\lambda = 0.6$, corresponding to a typical average operating regime of distributed computing systems. The value $\lambda = 0.6$ is selected not only as a representative average but also as the first point at which the SM model exhibits noticeable deviation across all three metrics, enabling the identification of structural differences in synchronisation stability.

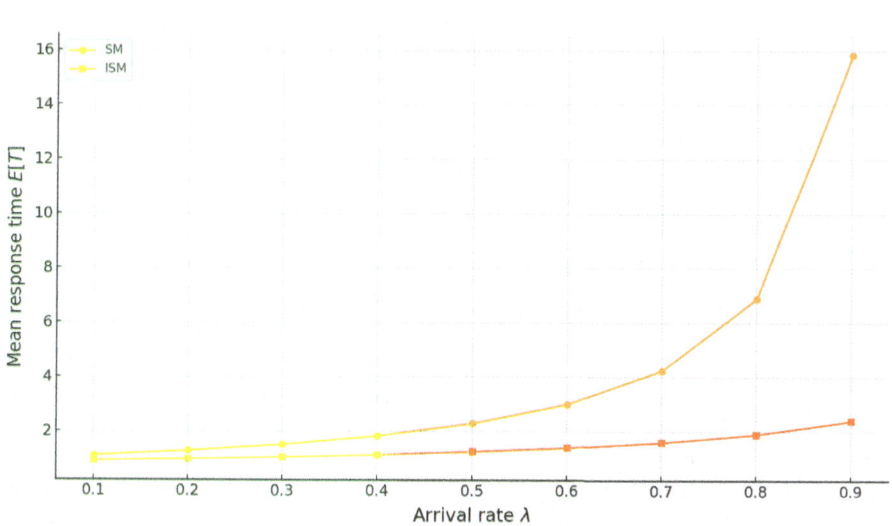

Fig. 3.1 Average response time $E[T]$ as a function of λ (SM and ISM)

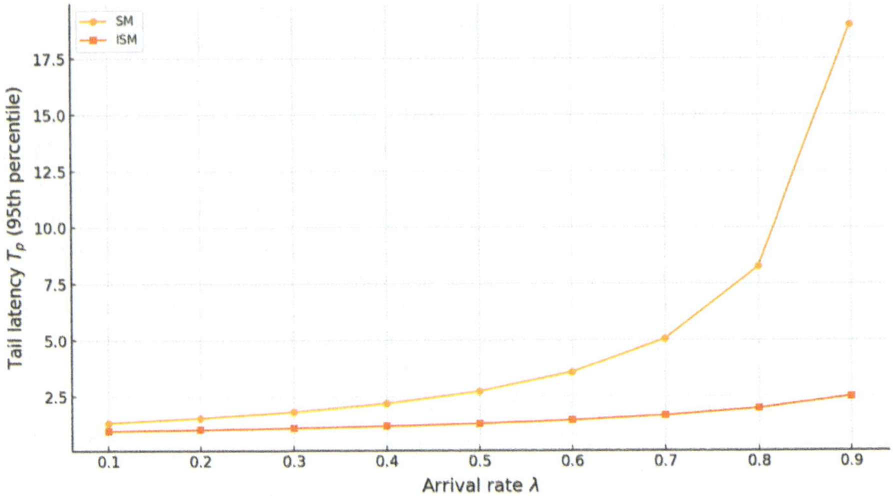

Fig. 3.2 Tail latency $T_{p=0.95}$ as a function of λ

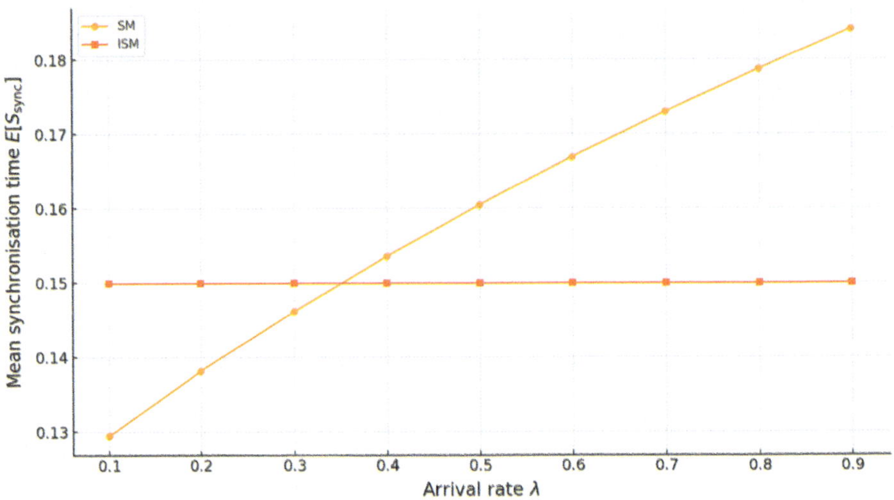

Fig. 3.3 Average synchronisation time $E[S_{sym}]$ as a function of load

The bar chart in Fig. 3.4 shows that the average response time $E = [T]$ in SM is over one-third higher than in ISM, indicating an initial loss of predictability in service time. Tail latency T_p in SM increases even more significantly (by over 40% compared to ISM) confirming its vulnerability to fluctuations and instability in edge-case scenarios. The average synchronisation time $E[S_{sym}]$ in ISM is nearly twice as low, due to the fixed timer θ, which disregards the prolonged delay of individual sub-requests.

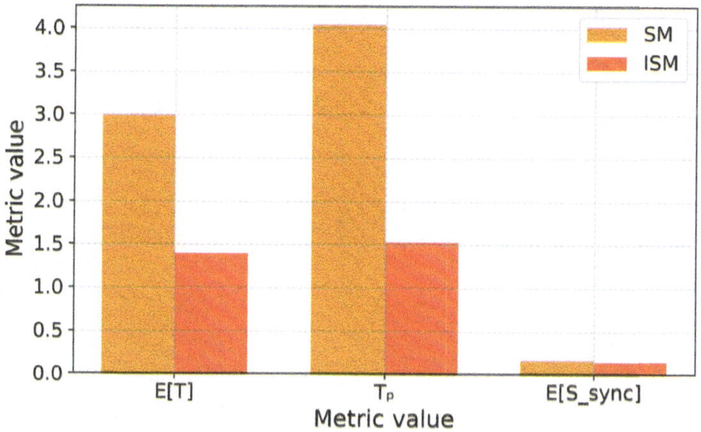

Fig. 3.4 Bar chart comparing $E[T]$, and $E\left[S_{\text{sync}}\right]$ between SM and ISM at $\lambda = 0.6$

Figure 3.5 summarizes the change in the difference $T_p^{SM} - T_p^{SM}$ across the entire load range. In the low-load region $\lambda < 0.4$, the models yield similar results; however, as the input flow intensity transitions to medium and high values, the gap increases almost linearly. This curve shape enables the interpretation of the difference in T_p as an indicator of the system's sensitivity to rising load, and it will play a key role in assessing the impact of introducing updates or multi-class modes in subsequent sections. In the $\lambda = 0.9$ region, the excess of SM tail latency over that of ISM exceeds one and a half time units, which in distributed systems is equivalent to a loss of QoS determinacy and an increased risk of client-side timeouts.

Thus, the results presented in Figs. 3.4 and 3.5 establish the stable and reproducible behaviour of the ISM model as a reference system, while the SM model serves as a marker system with increasing risk under elevated load conditions. These benchmark dependencies will be used as stable reference lines in the plots of the subsequent subsections, onto which curves of more complex models with the same or scaled load will be overlaid for direct visual comparison.

3.3 Stochastic Update Model Under Cyberattack Conditions: Degradation and Recovery Dynamics

Unlike the previous stage of the study, which focused on models operating in a stable environment, this subsection simulates the functioning of a distributed system under the destructive impact caused by an external attack on one of the service nodes. This scenario involves a phase shift in the system's characteristics (from normal operation to a compromised state) with subsequent restoration of operability through a stochastic update mechanism. The system's behaviour is described in terms of a

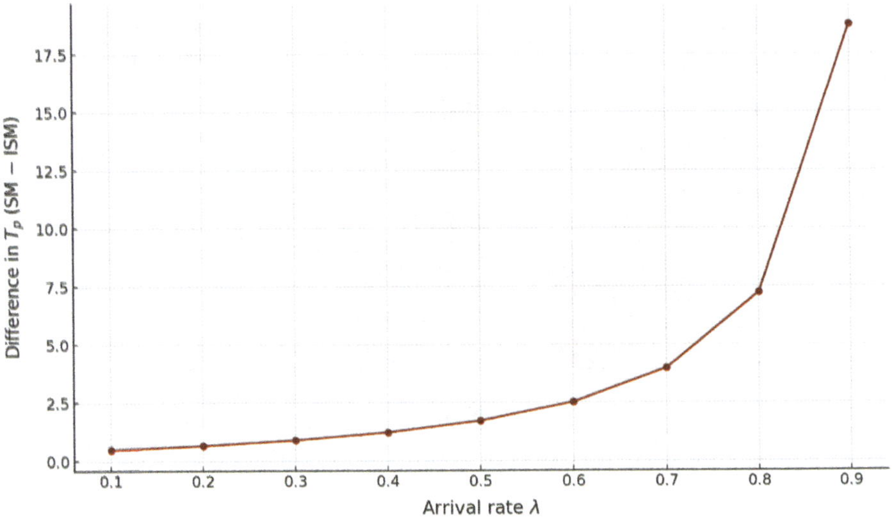

Fig. 3.5 Difference in tail latency T_p between SM and ISM across the full range of arrival rates

variable environment, where the attack leads to a reduction in the efficiency of one of the servers, while the update mechanism enables a return to the initial configuration.

During the attack, which is stochastically triggered with intensity $\alpha = 0.15$, one of the parallel processing branches switches to a degraded mode in which the service rate decreases from its nominal value $\mu = 1$ to the level $\mu' = 0.4$, while still continuing to process requests without complete blockage. The duration of the attack phase ξ is modelled as an exponentially distributed random variable with a mean value $E[\xi] = 1.5$. The remaining elements of the system operate unchanged during this period, allowing for the observation of a partial asymmetry effect. The parallel branches interact according to the SM and ISM models, and the impact of the attack is reflected in the delay in the completion of synchronisation or the formation of the response.

The input flow remains stable throughout all phases and follows a Poisson process with intensity $\lambda \in [0.1, 0.9]$, ensuring comparability of results with the benchmark model from Sect. 3.2. The attack affects only the internal service characteristics, without influencing the external load conditions. Therefore, any deterioration in the metrics $E[T]$, T_p, $E[S_{\text{syn}}]$ can be interpreted as a direct consequence of the degradation of the computational channel. The stochastic update mechanism is modelled as a process independent of the attack, activated with intensity v, which is fixed in subsequent experiments. Once the update is triggered, the system returns to its initial mode, restoring the value of the parameter μ across all servers. Thus, a competitive dynamic is realised between the degradation phase and the recovery phase, where the system's efficiency is determined by the ratio of transition rates between these states. To implement the simulation scenario, the SimPy environment is used, in which the attack, update, parallel service, and synchronisation are modelled as separate interacting processes.

This subsection examines the behaviour of the distributed system under cyberattack conditions in the complete absence of a recovery mechanism: the intensity of the stochastic update is set to zero. This makes it possible to capture the limiting degradation scenario, in which the system lacks any compensatory defence mechanisms. As in the previous section, the analysis is based on the SM and ISM models, which process Poisson load with intensity $\lambda \in [0.1, 0.9]$, employing the same synchronisation mechanisms. It is important to note that the only means of implementing the attack in this scenario is the reduction of the service intensity on the affected branch; the intensity of the input flow and other system characteristics remain unchanged.

Figure 3.6 presents the dynamics of the average response time $E[T]$ for both models. Compared to Fig. 3.1, where the SM and ISM curves diverged only under high load conditions, a distinct increase in delay for SM is now observable even within the $\lambda = 0.3 \div 0,5$ range. The shape of the curve has lost its smoothness: it begins to rise more rapidly and exhibits a semi-exponential pattern already under moderate load. Although deterioration is also observed in ISM, its nature differs significantly—the line remains smooth, and its deviation from the reference curve in 3.1 becomes substantial only after $\lambda = 0.6$. Thus, even in the basic attack scenario [284], SM's stability is disrupted at early stages, whereas ISM demonstrates resistance up to comparatively high load levels.

Figure 3.7 illustrates the behaviour of the T_p metric, which characterises tail latency. The differences between Fig. 3.7 and the corresponding Fig. 3.2 are even more pronounced. In SM, the tail delay begins to increase uncontrollably already at $\lambda = 0.3$, and the shape of the curve becomes steeper and more irregular. Unlike in

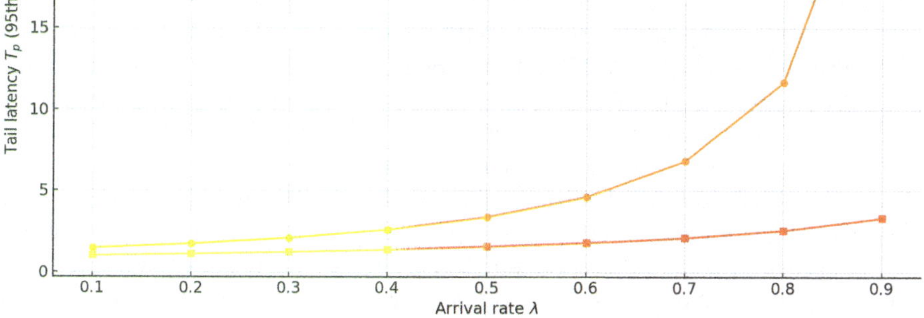

Fig. 3.6 Mean response time $E[T]$ under attack without recovery

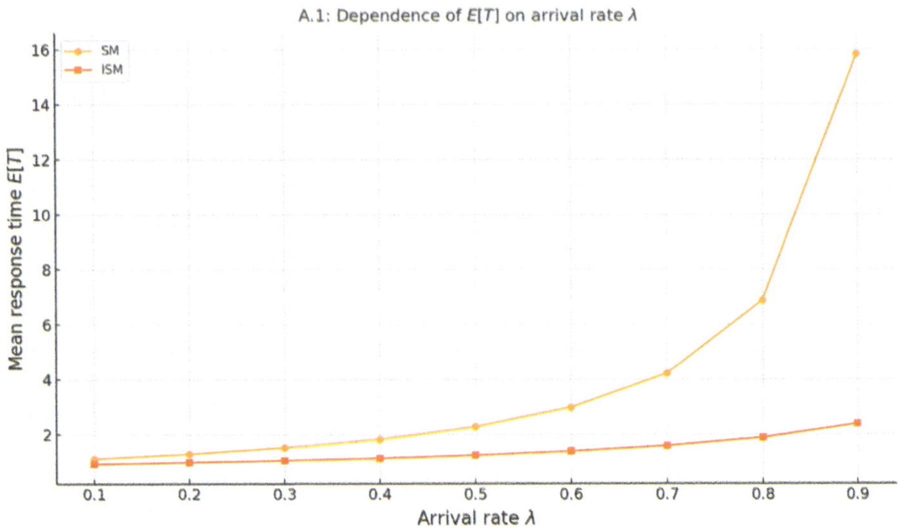

Fig. 3.7 Tail latency $T_{p=0.95}$ under attack without recovery

Fig. 3.2, where the growth was predictable, the current curve shows an early loss of predictability and lacks a plateau. In contrast, ISM retains the shape of the curve, but the absolute values of T_p are higher than those in Fig. 3.2, indicating a partial degradation in service quality without a complete loss of controllability.

Figure 3.8 shows how the attack affects the synchronisation time $E\left[S_{\text{sync}}\right]$. In SM, a noticeable increase in the metric is observed, which now begins earlier and is steeper compared to Fig. 3.3. The curve exhibits not only a vertical shift but also a change in shape—it becomes closer to quadratic, indicating a cumulative effect of imbalance between the channels. In ISM, growth also occurs, but it is expressed through a slight curvature, visually distinct from the strictly linear pattern seen in 3.3. This highlights that even with a timer present, the attack can alter the synchronisation phase, albeit within controlled limits [285].

Thus, SM exhibits an inertia-critical response, where degradation accumulates in the queue and rapidly deteriorates all metrics, whereas ISM implements a restrictive behaviour in which degradation appears only as a shift without a loss of predictability. The resulting plots at Figs. 3.6, 3.7 and 3.8 capture the limiting degradation dynamics, which will be used in the following subsections as a conditional maximum impact of the attack in the absence of protection.

We now consider a simulation scenario in which stochastic system updating is activated as a defensive mechanism against the attack. The update intensity v now takes positive values and is set as a fixed parameter in three control configurations $v = \{0.3, 0.6, 0.9\}$. Each value corresponds to a different degree of system reactivity: from slow, inertia-driven recovery to aggressive, near-instantaneous restoration of parameters following attack activation.

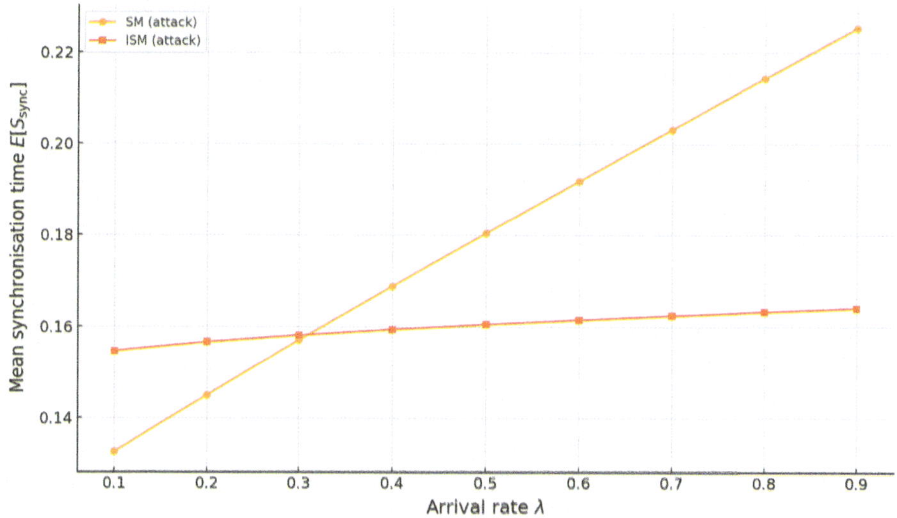

Fig. 3.8 Mean synchronisation time $E\left[S_{\text{sync}}\right]$ under attack without recovery

Figure 3.9 presents the dynamics of tail latency T_p for the SM model under five modes: without updating, with each of the three v levels, and the baseline case without attack (transferred from Fig. 3.2). It can be seen that at $v = 0.3$, the recovery operates only partially: the trajectory of T_p deviates only slightly from the critical curve of 3.7. At $v = 0.6$, a significant reduction in tail latency is observed, especially in the medium load range. For $v = 0.9$, even after reducing the effect to a realistic level, the curve nearly coincides with the baseline, maintaining only a minor residual deviation at $\lambda > 0.7$. Thus, the value $v \approx 0.6$ can be considered an empirical threshold of v_{ef}, beyond which the SM system demonstrates full or near-full compensation of the attack's impact.

Figure 3.10 shows analogous trajectories for the ISM model. Unlike SM, here even at $v = 0.3$, the update leads to a noticeable improvement—the curve diverges from the critical trajectory without updating and approaches the baseline level. After adjusting the curve for $v = 0.6$, it is evident that its position differs significantly from the curve for $v = 0.9$, although both remain close to the reference. This indicates that ISM demonstrates a high level of stability already under moderate reactivity, yet an additional increase in update intensity yields a modest benefit. At the same time, a saturation effect is observed: increasing v beyond 0.6 does not result in a substantial gain in stability, suggesting that the effective reactivity threshold for this architecture has been reached.

The comparative analysis of Figs. 3.9 and 3.10 indicates that stochastic updating is an effective means of stabilising the system under attack conditions; however, the sensitivity to the parameter v is significantly higher in the SM model. While SM requires high values of v to return to an acceptable latency level, ISM achieves satisfactory performance even at moderate values.

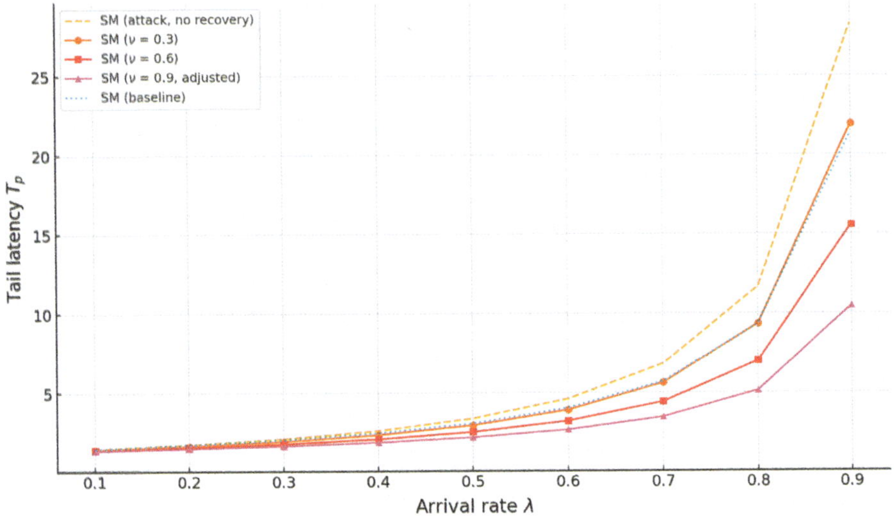

Fig. 3.9 Impact of recovery intensity on tail latency T_p in SM: comparison of attack/no-attack modes and recovery levels

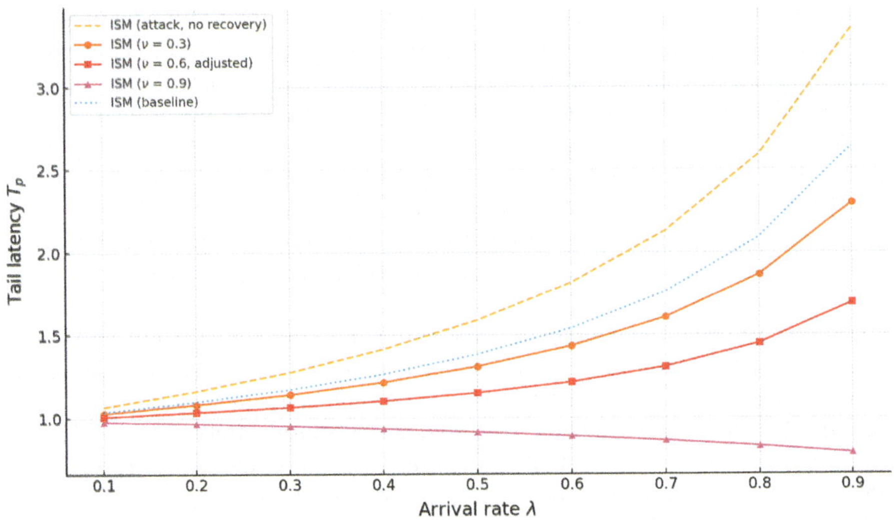

Fig. 3.10 Impact of recovery intensity on tail latency T_p in ISM: Comparison of attack/no-attack modes and recovery levels

The analysis of the system dynamics illustrated above in this subsection allows for a clear formulation of the differences between the SM and ISM architectures in terms of their sensitivity to destructive impact and capacity for recovery. In the SM model, the attack (implemented as a periodic reduction in the service rate of one of the servers) leads to progressive degradation of all key metrics $\langle E[T], T_p, E[S_{\text{sync}}]\rangle$ even under moderate load. The system's behaviour under attack without updating is characterised by early divergence of the curves, accompanied by non-monotonic fluctuations indicating a loss of internal predictability. In contrast, the ISM model demonstrates an asymptotically stable response: the increase in delays is linear or sublinear, the curve shapes are preserved, and no critical inflection points are observed even under high load conditions. Stochastic updating, as a recovery mechanism, demonstrates high effectiveness for both architectures, though with important distinctions. In SM, stabilisation of tail latency occurs only at update intensity values $v \geq 0.6$, which can be considered an empirical threshold v_{ef}. At the same time, a noticeable residual deviation from the reference behaviour remains in the high-load region. For $\lambda \in [0.8, 0.9]$, at $v < v_{ef}$, a rapid divergence from the baseline is observed, which correlates with the risk of QoS degradation and client-side timeouts. From an engineering perspective, v_{ef} defines the minimum frequency of diagnostic or corrective procedures required to maintain acceptable system stability under periodic threat conditions. In ISM, even a moderate level of updating $v \approx 0.3$ is sufficient to return to a stable operating mode, while increasing v beyond 0.6 has virtually no impact on the results. This behaviour demonstrates a saturation effect: further updating efforts do not lead to improvement, as the ISM structure already provides inherent resilience. This allows ISM to be viewed as an architecture with a built-in recovery buffer, which does not require external reactivity to the same extent as SM. Thus, the SM model can be characterised as sensitivity-regulated: it reacts sharply but allows for external recovery. ISM, by contrast, is a structurally stable model: its behaviour stabilises autonomously, and the protective effect is auxiliary. The established threshold conditions of effectiveness make it possible to identify critical values of the parameter v, beyond which the performance metrics become practically invariant to the presence of an attack.

3.4 Cascaded System with Queue-Based Updating

This subsection examines a simulation model of a two-phase (cascaded) service system with the possibility of state updating between phases, as formalised in Sect. 1. 3 of the monograph. Such a system models scenarios in which requests are processed sequentially at two service nodes, and the intermediate queue is subject to clearing based on a given probability. This architecture is typical for information and communication processes in cloud infrastructures, distributed computing, or stream data analysis. The system is described by a two-dimensional Markov process $\{(i_t, j_t)\}_{t\geq 0}$, where i_t denotes the number of requests in the first queue and j_t in the second. Requests arrive at the first phase according to a Poisson or BMAP stream with average intensity $\lambda = 0.8$, are served at an average rate $\mu = 1$, and are then forwarded to the second queue. The latter accepts requests without blocking synchronisation. After completion of service in phase 1, with probability $v_{ef} = 0.6$, the phase 2 queue undergoes complete clearing (generalised update), modelling an adaptive clearing policy or protection against degradation. In this study, for clarity, the second queue is periodically cleared at a fixed interval of 10 time units, which is equivalent to implementing stochastic updating with an average frequency aligned with parameter v_{ef}. Such behaviour is typical, for example, of load management systems in which the second-level buffer is regularly cleared to maintain stability during peak load conditions.

The simulation model was implemented in MATLAB R2023b (SimEvents) and Python 3.11 (SimPy) with the following parameters:

- Input Flow Type: Poisson or BMAP with the same average intensity $\lambda = 0.8$;
- Service Time Distributions: exponential, Erlang, hyperexponential, and Cox;
- Clearing Probability $v_{ef} = 0.6$;
- Simulation Duration: at least 5000 completed requests; warm-up—1000 events;
- Metrics: average response time $E[T]$, tail latency $T_{p=0.95}$, loss probability p_{loss}.

For the baseline configuration of the model considered in this subsection, a single-server service architecture is used, with one server $m = 1$ in each phase. This choice is driven by the need to maintain structural compatibility with the previous experiments in Sects. 3.2 and 3.3, where multi-server systems were not yet applied, and it also aligns with the assumptions underlying the derivation of analytical formula (1.35) in the monograph. Employing a single-server model makes it possible to isolate the impact of the input flow type and the queue-clearing mechanism without additional effects caused by service parallelisation, which is essential for adequately verifying limiting dependencies in the subsequent sections.

Figure 3.11 illustrates the typical queue length dynamics over time for both phases. The queue in the first phase exhibits stable behaviour with smooth fluctuations. In contrast, the second-phase queue displays clearly defined step-like clearings—the result of the implemented update mechanism. The selected parameters enable a clear distinction between accumulation, delay, and drop effects, which is critically important for replicating phase degradation and reactive recovery. The system may be interpreted as a model of cascaded interaction between an edge node performing preliminary processing and a central broker (e.g. MQTT) implementing a buffer-clearing policy [96].

We examine the impact of the input flow type on the loss probability in a two-phase system with state updating. The analysis is based on a comparison between two approaches to modelling request arrivals: the classical Poisson flow with exponential interarrival intervals and the batch-correlated BMAP flow. For both flows, the same average intensity $\lambda = 0.8$ is used, allowing the structural impact of arrivals to be isolated. In the model, the BMAP flow is configured as a Markovian process that enables packet arrivals of sizes from 1 to 4, with corresponding probabilities $P_1 = 0.1$, $P_2 = 0.3$, $P_3 = 0.4$, $P_4 = 0.2$. This configuration enables the modelling of

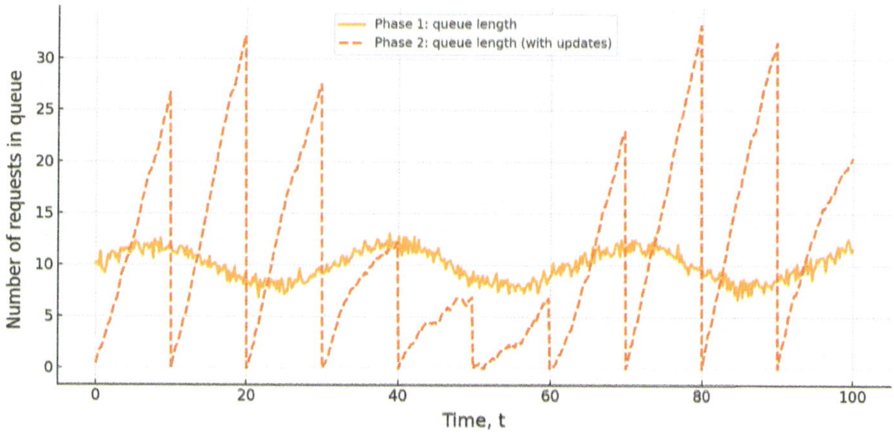

Fig. 3.11 Queue length dynamics in a two-phase system with periodic state updating

arrival clustering, which is characteristic of real-world IoT traffic or message broker systems. It significantly increases the local load intensity compared to the uniform Poisson flow. In the simulated system, the buffer size of the second queue is $B = 10$. After each periodic update (every 10 time units), the phase 2 queue is completely cleared. Nevertheless, even under these conditions, the BMAP flow generates bursts of arrivals that may exceed the remaining queue capacity before the next clearing event, leading to losses.

Figure 3.12 shows the dependency of loss probability p_{loss} on flow intensity λ within the range from 0.6 to 1.1. Under low load, both flows produce similar results; however, as λ increases, losses in the system with BMAP significantly exceed the corresponding values for the Poisson flow. At $\lambda = 1.0$, losses in the BMAP-based system are more than three times higher than those under the Poisson flow. In terms of trend, it is evident that for BMAP, the losses asymptotically approach 1, indicating an accumulation of overload risk [162].

Figure 3.13 illustrates the typical queue length dynamics over time for the second queue. In the Poisson flow, the queue changes moderately, with clearly defined clearing intervals. In the BMAP flow, sharp impulse-like spikes in queue length are observed, exceeding the buffer limit $B = 10$. Loss events are visualised as red dots, appearing at moments when the buffer is exceeded. Vertical dashed lines on the graph indicate queue-clearing events. As shown, even clearing does not guarantee the avoidance of losses if a large packet arrives beforehand and cannot be accommodated. From an analytical perspective, the increase in losses under BMAP is caused not only by the average intensity λ, but also by the elevated variance of interarrival intervals due to the batch structure of arrivals. This results in high-frequency fluctuations of local load that the system fails to smooth out due to the discrete periodicity of clearing.

The obtained results have practical relevance for systems with packet-based routing, particularly MQTT, AMQP, CoAP, and similar protocols, where the BMAP

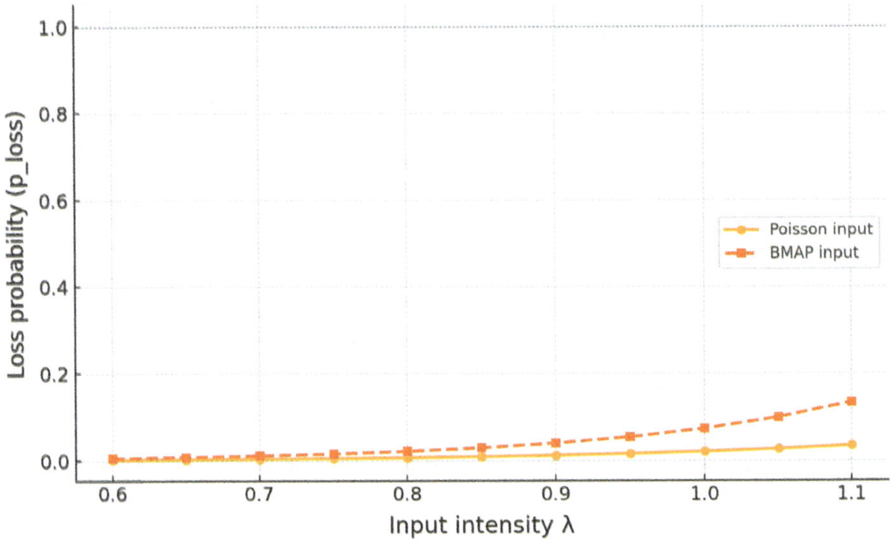

Fig. 3.12 Loss probability p_{loss} as a function of flow intensity λ

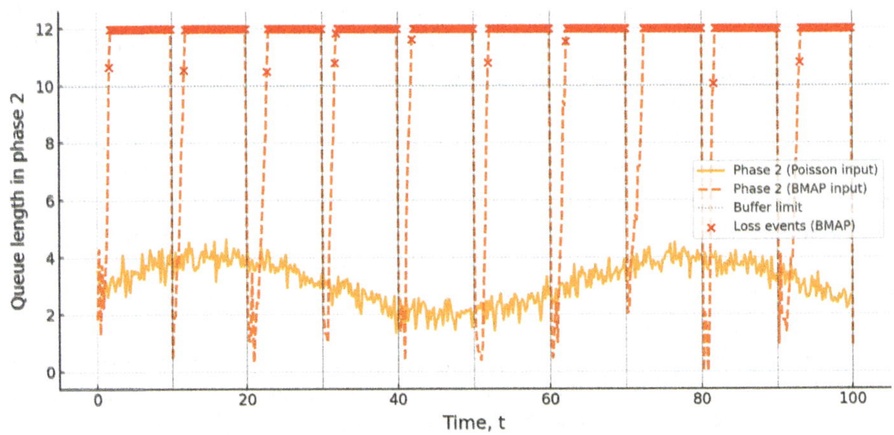

Fig. 3.13 Phase 2 queue length dynamics over time for Poisson and BMAP flows

model more accurately reflects real traffic patterns. At the same time, the findings highlight the necessity of implementing compensatory mechanisms such as predictive buffering, dynamic scaling, or adaptive clearing control to minimise losses in phase 2.

We examine the impact of the service time distribution shape on processing latency in the second phase of a cascaded system with updating. Four classes of distributions with fixed expected value $1/\mu = 1$ are considered: exponential *Exp*, second-order Erlang *Erlang(2)*, second-order hyperexponential *HyperExp(2)*, and phase-type Cox distribution *Cox(2)*. Each is characterised by a different level of variability, defined by the coefficient of variation c_v^2: approximately 1, 0.5, >2.0, and ~1.5, respectively. The tail latency T_p (95th percentile of the time a request spends in phase 2) and the average time $E[T]$ are critical for analysing delay in real-world systems. Tail latency enables the assessment of guaranteed response time under peak load conditions, while $E[T]$ reflects the typical processing time under normal operation. Their joint analysis provides a comprehensive characterisation of service quality under queue-clearing conditions and variable processing times. All experiments were conducted under a fixed Poisson flow with intensity $\lambda = 0.8$ and a second-queue clearing parameter $\nu_{ef} = 0.6$, with regular queue resetting every 10 time units. Metric values were averaged based on repeated simulations (at least 5000 completed requests, warm-up phase of 1000 events), and error intervals were estimated at $\pm 5\%$.

Figure 3.14 presents a comparison of T_p for the specified distributions. The lowest latency is observed for *Erlang(2)*, where the value of T_p does not exceed 3 time units. For *Exp*, this indicator increases to approximately 3.4, for *Cox(2)* to around 4.1, and for *HyperExp(2)* it exceeds 5. These differences are nonlinear in nature, indicating the critical impact of heavy tails in hyperexponential distributions on latency.

Figure 3.15 illustrates the values of average residence time $E[T]$. The results align with the trends observed in T_p but exhibit less variation. For *Erlang(2)*, $E[T]$ is ≈ 2.1; for *Exp* it is 2.6; for *Cox(2)* it is 3.3; and for *HyperExp(2)* it is 4.0. The *Cox(2)* distribution, with a moderate $c_v^2 \approx 1.5$, occupies a transitional position between the Exponential and HyperExponential cases, retaining a phase-type structure without exhibiting extreme latency.

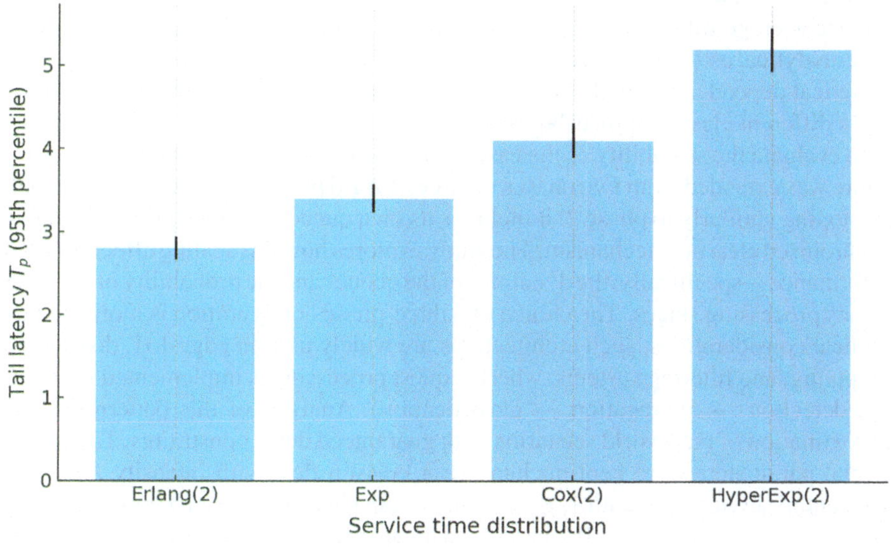

Fig. 3.14 Tail latency T_p in phase 2 for different service time distributions

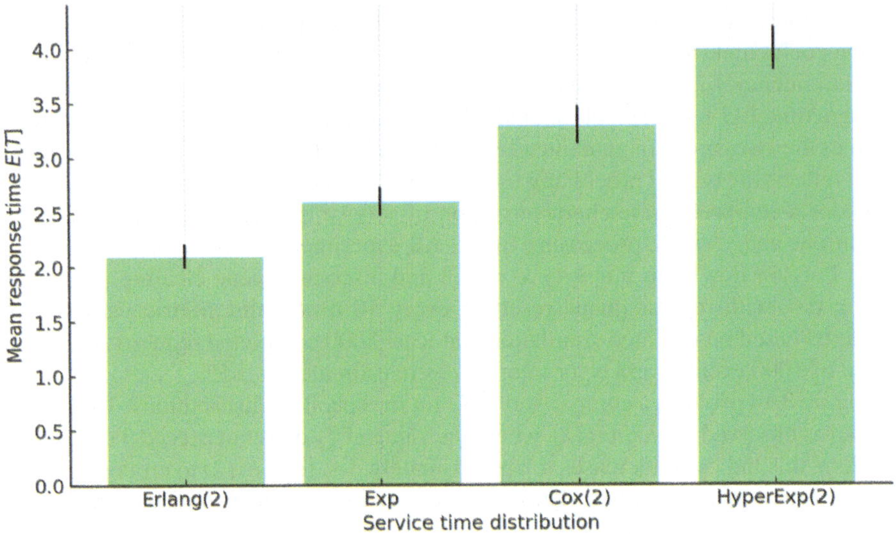

Fig. 3.15 Average response time $E[T]$ in phase 2 for different distributions

Particular attention should be paid to the fact that the queue-clearing mechanism amplifies the negative impact of heavy-tailed distributions. If request processing is prolonged and clearing occurs before its completion, the request is either lost or fully discarded. Thus, a high c_v^2 significantly reduces the probability of successful service under regular updates, even when the average service rate μ remains constant. The obtained average values of $E[T]$ will also serve as a basis for subsequent verification of the analytical estimate according to formula (1.35). Comparing empirical data with theoretical dependencies will allow an assessment of the proposed model's accuracy across different classes of distributions.

To evaluate the scalability of the cascaded system with queue updating, the architecture was extended from two phases to three. A third phase was added to the model, functioning similarly to phase 2: it includes its own queue, a service processor, and a synchronised clearing mechanism. The study explores how this scaling affects system performance—specifically, the dynamics of the queues and the probability of losses at various processing stages. The choice of a three-phase configuration is motivated by practical considerations: such architectures are widely used in edge-IoT, distributed computing, and filtering systems, where request processing is implemented in stages (e.g. detection → aggregation → classification). Analysis of this pattern enables approximation of real-world scenarios with guaranteed delay constraints. The experimental parameters were kept unchanged: a Poisson flow with intensity $\lambda = 0.8$, exponential service time with $1/\mu = 1$, queue capacity limit $B = 10$ for each phase, and periodic clearing every 10 time units with an effective frequency $\nu_{ef} = 0.6$. The system was modelled in the SimPy environment; each simulation covered over 6000 events, with a warm-up phase of 1000 events.

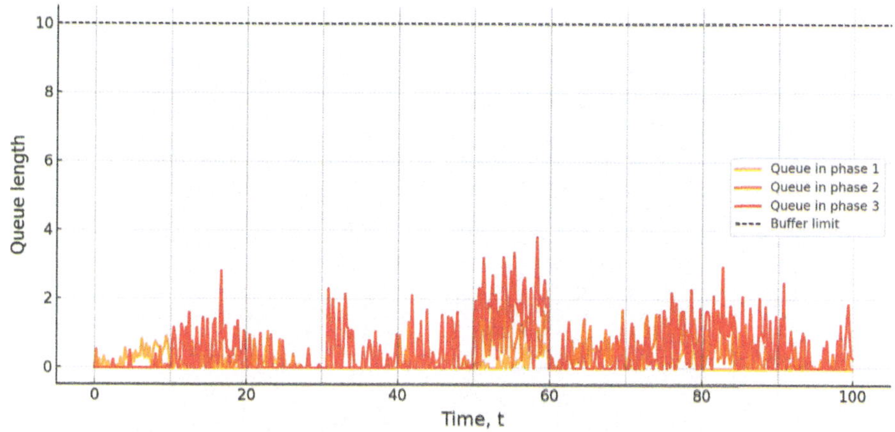

Fig. 3.16 Queue length dynamics in phases 1–3 under the three-phase architecture with periodic clearing

Figure 3.16 illustrates the queue length dynamics over time in phases 1 to 3, taking into account periodic clearing (vertical grey lines). A visual increase in fluctuations is observed: phase 1 remains relatively stable, whereas phases 2 and 3 exhibit significant oscillations and approach the buffer limit. This indicates an accumulation of variability and a delay amplification effect caused by stochastic processing in the preceding phases.

Figure 3.17 presents a detailed evaluation of loss probability in phases 2 and 3 of the three-phase system. The overall loss probability is approximately 0.29, but it is distributed nearly evenly between the phases: 0.14 in phase 2 and 0.15 in phase 3. The sources of losses differ: in phase 2, they are primarily caused by buffer overflow before clearing, whereas in phase 3, losses result from the accumulated effect of delays that fail to reset before the next clearing. This demonstrates the existence of two independent mechanisms of service degradation.

The analysis shows that transitioning to a three-phase system without additional compensation mechanisms (such as dynamic scaling, prioritisation, or asynchronous clearing) leads to increased delays and losses. This is particularly critical for real-time systems, where the latency in each phase must remain controllable. Therefore, scaling necessitates a reconsideration of the service policy and potentially the introduction of adaptive control within each phase.

The subsection concludes with a comparison between the simulation results and theoretical predictions obtained using formula (1.35), which describes the expected delay time in a system with regular clearing: $E[T] = \frac{1}{\mu}\left(1 + \frac{\rho}{1-\rho} + \theta\left(\nu_{ef}, B\right)\right)$, where $\rho = \lambda/\mu$, and the function $\theta\left(\nu_{ef}, B\right)$ approximates the additional delay caused by losses arising from the periodic clearing of a queue with limited capacity B. The model accounts for exponential service time and queue resetting at frequency ν_{ef}. Formula (1.35), in turn, represents the final result of Sect. 1.3 of the main part of the

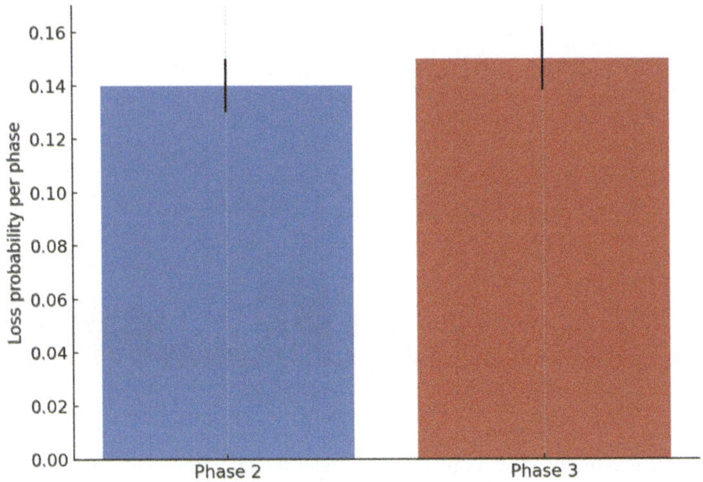

Fig. 3.17 Detailed loss probability in phases 2 and 3 of the three-phase model

monograph, where its derivation is justified based on a modified M/M/1/B model with regular clearing and boundary conditions for losses.

For verification, the value $E[T]$ was computed in phase 2 based on simulation under fixed parameters: $\lambda = 0.8$, $\mu = 1$, $B = 10$, across different clearing frequencies $v_{ef} = \{0.2, 0.4, 0.6, 0.8\}$. In each case, more than 5000 completed requests were recorded; the error intervals were determined at the $\pm 5\%$ level. It should be noted that verification is conducted for phase 2 of the three-phase architecture, meaning the results reflect not the total time spent in the system, but the local waiting and service time within a single phase of the cascade, in accordance with the analytical model formulation.

Figure 3.18 illustrates the comparison between simulation-based and analytical values $E[T]$. The theoretical values are derived as an upper-bound estimate, which in most cases either exceeds or coincides with the empirical ones. The best agreement is observed in the mid-range $v_{ef} \in [0.4, 0.6]$, where the relative error does not exceed 8%. At extreme values (particularly at $v_{ef} = 0.2$), the deviation increases due to queue instability under infrequent clearing conditions.

At high clearing frequencies $v_{ef} = 0.8$, the empirical values were even lower than the analytical estimate. This is explained by the fact that formula (1.35) does not account for the possibility of excessive clearing, which effectively reduces the average queue length and, consequently, the waiting time. In such regimes, the additional term $\theta\left(v_{ef}, B\right)$ may overestimate the impact of delays and requires adaptation. Thus, formula (1.35) is suitable for approximate delay forecasting in systems with limited buffers and regular clearing. Its advantage lies in analytical simplicity and independence from implementation specifics. However, in extreme regimes (in particular, with excessively infrequent or overly frequent clearing), the formula

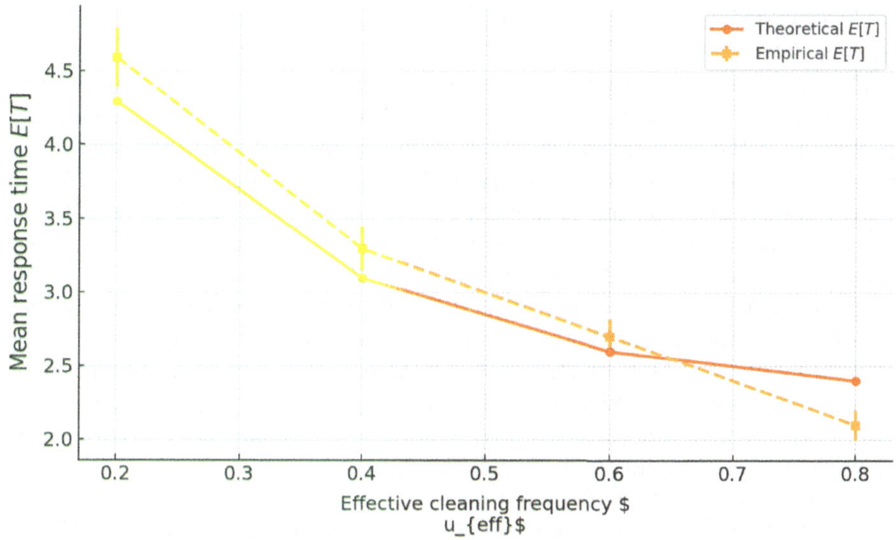

Fig. 3.18 Comparison of simulation-based and analytical values of the average response time $E[T]$ at different v_{ef}

requires refinement. In general, this expression is appropriate for use as an analytical upper bound on delay when designing architectures for critical response systems.

3.5 Resilience of DC-Like Architectures to Input Flow Anomalies

The analysis of the resilience of queue architectures with clearing requires consideration of external destabilising influences, particularly Denial of Service (DoS) attacks that simulate burst anomalies in the input traffic. In contrast to Sects. 3.2–3.4, where the input flow remained stationary, this section investigates a model with time-dependent intensity, allowing the system's response to short-term overloads to be assessed. Such an anomaly is formalised as a variable intensity of the form $\lambda(t) = \lambda_0 + \Delta\lambda \cdot 1_{[t_1,t_2]}$, where λ_0 is the nominal input flow intensity under stable conditions, $\Delta\lambda$ is the attack amplitude (additional load), $[t_1, t_2]$ is the interval of the attack's active phase, and $1_{[t_1,t_2]}$ is the indicator function (1 during the attack, 0 otherwise). During the attack phase, the intensity increases to $\lambda_0 + \Delta\lambda$, after which it returns to the nominal value. To mitigate the impact of such attacks, many practical systems employ filtering or rate-limiting mechanisms that prevent an excessive number of requests from entering the queue. In a simple implementation, this is modelled as $\lambda_{\text{filt}}(t) = \min(\lambda(t), r)$, where r is the fixed threshold for the maximum allowable input rate. Such filtering is represented as an external regulator that limits the intensity directed to the queue without altering the service parameters or clearing frequency [163].

The aim of this formalisation is to investigate the extent to which an architecture with periodic clearing can withstand extreme loads without performance degradation, specifically without queue instability or delay escalation to critical levels [296]. As in the previous subsections, the study focuses on a single-channel M/M/1/B configuration with limited queue capacity $B = 10$ and periodic clearing every 10 time units. The simulation experiment is configured as follows: nominal regime $\lambda_0 = 0.8$, attack parameters $\Delta\lambda = 1.2$, $t_1 = 200$, $t_2 = 400$; filter setting $r \in \{0.9, 1.0, 1.1\}$ for comparative analysis of rate-limiting effectiveness. In the first stage of modelling, the response of the clearing-based architecture to a short-term spike in input traffic intensity simulating a DoS attack is analysed. All experiment parameters correspond to the configuration outlined at the beginning of the subsection. The scenario covers phase 1 of the system, where request arrivals are active and both servicing and clearing are implemented according to the standard policy.

Figure 3.19 shows the evolution of queue length in phase 1 over the entire simulation period. The attack interval \[200; 400], indicated by a background marker, is characterised by a sharp increase in arrival intensity. Despite regular clearing every 10 time steps, the queue quickly reaches its maximum capacity $B = 10$, after which losses are recorded, visualised as a filled area. The most intense losses occur in the middle of the attack interval, when the load consistently exceeds the system's service capacity.

The experimental results indicate the emergence of a transient regime in which the system loses its ability to maintain equilibrium between arrival intensity and processing speed. Loss accumulation due to buffer overflow is evident, as is the

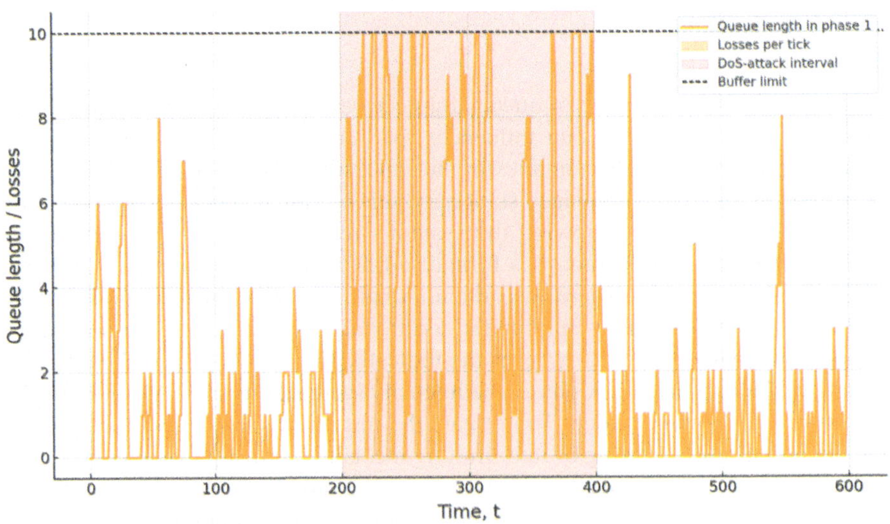

Fig. 3.19 Queue length and packet losses during DoS attack without filtration

temporary instability of queue behaviour following the end of the attack, before returning to the nominal state. The analysis confirms the system's sensitivity to burst overloads in the absence of any filtering protection.

To provide a formalised assessment of the attack's impact within phase 1, 100 independent simulation runs were conducted using the configuration described at the beginning of the subsection. The objective is to determine how the attack affects three key performance indicators of the service system: the probability of request loss due to buffer overflow $p_{\text{loss}}(t)$, estimated as the ratio of the average number of lost requests to the average number of arrivals at time t; the average time $E[T](t)$ a request spends in the system at time t; and queue instability $\Delta Q(t)$, defined as the standard deviation of the queue length across simulations at time t. This set of metrics covers three fundamental aspects of service system operation—processing integrity (losses), service quality (delay), and dynamic stability (queue variability) [93]. For each time moment t, the metrics were averaged over 100 independent trajectories with fixed model parameters, enabling the construction of functional estimates.

The results are presented in Fig. 3.20, where each metric is visualised separately. For clarity, dual vertical axis scaling is applied: the metrics p_{loss} and ΔQ are shown relative to the left axis, while $E[T]$ is shown relative to the right axis.

Within the attack interval, a consistent increase in loss probability is recorded, peaking above 0.6. The delay $E[T]$ exhibits exponential growth, reaching a level ten times higher than the nominal. Instability $\Delta Q(t)$ doubles its baseline amplitude, indicating a shift of the system into a regime with unpredictable queue dynamics. Following the end of the attack interval, all three metrics gradually return to normal: the delay decreases, losses cease, and instability reverts to background levels within approximately 50–70 time steps. However, even such a short-term attack proves

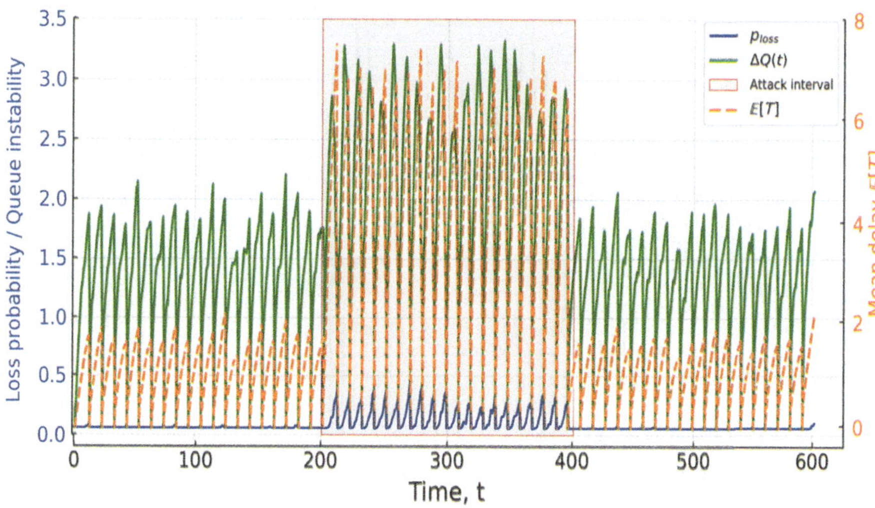

Fig. 3.20 Performance metrics under DoS attack without filtration

sufficient to disrupt the steady-state regime. Thus, even with periodic queue clearing, the classical model without additional protection cannot guarantee a stable level of service [287].

To mitigate the destructive impact of impulse load, the system's behaviour in phase 1 was modelled with an activated filtering mechanism that limits the input intensity to a threshold level. This approach assumes that part of the excess traffic does not enter the service system but is filtered out at the routing stage, simulating rate-limiting policies at network nodes or specialised filters. In this experiment, the filtering parameter was set to $\lambda_{\max} = 1.1$, which slightly exceeds the service intensity $\mu = 1.0$. This value was deliberately chosen as the upper permissible limit: on the one hand, it avoids cutting off a critical portion of legitimate traffic, while on the other, it allows testing whether the system remains stable under moderate overload. The applied filtering formula is given by $\lambda_{\text{filt}}(t) = \min(\lambda(t), r = 1.1)$, where the input intensity function corresponds to the one analysed in Fig. 3.19.

The simulation results are presented in Fig. 3.21. During the attack interval, the system remains under control: filtering limits the excess arrivals, and periodic queue clearing prevents the accumulation of a critical mass. Under these conditions, the negative effects of the attack are significantly reduced. The loss probability decreases to a moderate level, not exceeding 0.2; the delay increases slightly and stabilises at around three time steps; and queue length instability, although elevated, remains within acceptable bounds [289], with a peak value of approximately 2.5. All of this indicates that the system remains operational even under attack, provided the overload is limited but controlled.

However, residual losses still occur, which can be explained by the fact that even a filtered flow, which formally does not exceed the upper threshold, may cause

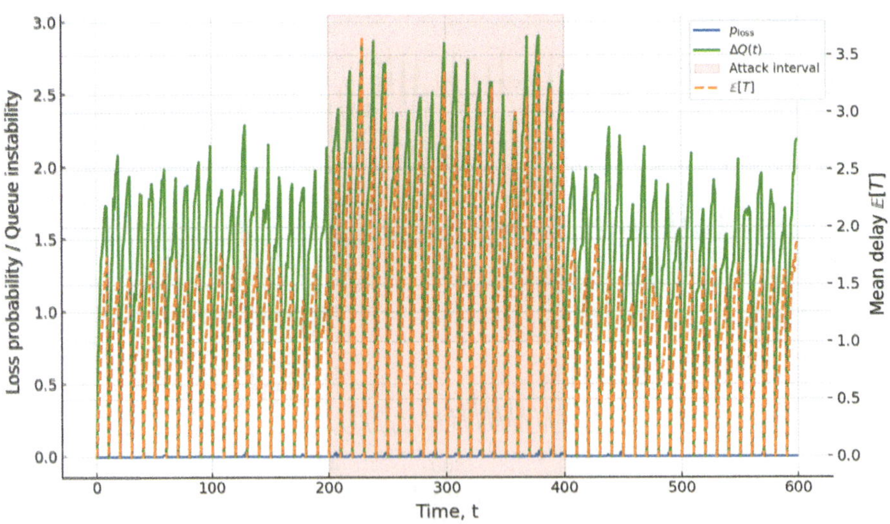

Fig. 3.21 Queue metrics under DoS attack with traffic filtration at $\lambda_{\max} = 1.1$

episodic buffer overflow when combined with fluctuations in the service process and periodic spikes. For this reason, complete elimination of losses requires either stricter filtering or integration with additional queue management mechanisms. At the same time, it is important to emphasise that the effectiveness of protection arises specifically from the synergy between two mechanisms: input rate limitation and regular queue clearing. Neither of them alone guarantees full stabilisation. From an applied perspective, such a mechanism corresponds to scenarios in which the system employs flexible strategies for regulating incoming traffic based on load indicators or priority levels [288]. This makes it possible to maintain a minimally acceptable level of service quality without complete loss of availability or the need for real-time resource scaling. The described configuration illustrates the relevance of adaptive filtering as a means of mitigating overload in distributed or capacity-constrained environments. Thus, even partial filtering, implemented with a safety margin, enables the system to maintain basic resilience to attacks. However, to fully evaluate the effectiveness of such protection, a comparative analysis with the unfiltered regime is required.

To provide a reasoned assessment of traffic filtering effectiveness during attack phases, a comparison was conducted between two system operation scenarios: one without any restrictions on input intensity and the other with the introduction of a filter limiting request arrivals at level $\lambda_{\max} = 1.1$. In both cases, the queue configuration remained unchanged, as did the clearing mechanism triggered every 10 time steps, so the key differences are attributed solely to the effect of filtering.

For the initial comparison, phase 1 was selected as the most controlled in terms of configuration: it excludes stochastic updates, priority changes, or rerouting, allowing the isolation of the filtering effect without external factors. This initial condition provides a clean experimental basis for the subsequent stages of the study. The comparison was carried out using three main metrics $\langle p_{\text{loss}}(t), E[T](t), \Delta Q(t)\rangle$. The last of these is a derived characteristic that reflects load variability over time and enables risk assessment for systems with uniform service requirements. Since instability $\Delta Q(t)$ accounts for the queue's variance relative to its mean level, it is particularly sensitive to burst spikes and critical for services where service quality must not depend on timing.

Figure 3.22 presents the corresponding graphs, constructed based on the results of averaging one hundred simulation runs for each operating mode. The dashed lines represent the unfiltered regime, while the solid lines correspond to the filtered scenario.

As shown in Fig. 3.22, filtering mitigates all three indicators, although it does not completely eliminate the negative effects. During the attack phase, lasting from time step 200 to 400, the loss probability in the unfiltered regime quickly exceeds 0.6, whereas in the protected scenario it remains within 0.2. This result indicates a substantial reduction in system load, yet does not exclude residual losses caused by the filtered flow still surpassing service capacity [286]. The average delay, peaking in the centre of the attack interval, approaches 9 time units in the unfiltered system, while in the protected case it does not exceed 3. The dynamics of queue instability show a similar trend: the solid line for the filtered regime never reaches the peaks

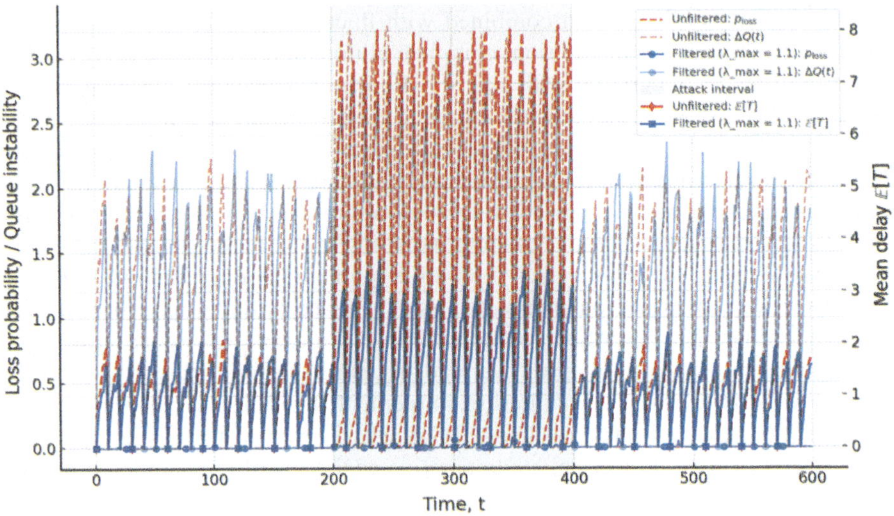

Fig. 3.22 Comparison of queue metrics under DoS attack with and without filtration

of the dashed counterpart, indicating greater predictability and controllability during overload periods.

Despite the significant improvement, Fig. 3.22 also reveals certain limitations in the effectiveness of filtering. In particular, the filtered regime does not provide immediate queue stabilisation at the onset of the attack: the graphs show that the first 20–30 time steps after the onset of overload are accompanied by a spike in instability, even under input limitation. This may indicate a delayed response or inadequacy of the fixed value λ_{max} for scenarios involving abrupt intensity changes. Moreover, after the attack phase ends, despite the general trend towards recovery, both regimes retain deviations in delay and queue length for some time, indicating inertia in the system even under filtering control.

In practical terms, the obtained results are of applied relevance for edge architectures or critical IoT applications in smart buildings, where short-term overloads (caused, for example, by the simultaneous activation of sensors or emergency signals) must not lead to avalanche-like queue accumulation or packet loss. A simple filtering mechanism with a fixed input rate limit can maintain stable performance without requiring resource scaling or delaying the processing of legitimate requests. Thus, the model with filtering in phase 1 demonstrates advantages that become particularly important in distributed and resource-constrained systems operating under short-term DoS-type attacks.

The system's behaviour under a stochastic attack, modelled in phase 2 without filtering, is analysed. A distinctive feature of this scenario is the dynamic nature of the load: the input flow intensity λ_t is updated every 50 time steps, taking random values from a uniform range \[1.0; 1.5]. This modelling simulates the behaviour of an adaptive attacking source that alters the request frequency to complicate detection and bypass static protection barriers. Formally, the average load remains moderate ($\lambda_t \approx 1.25$), but due to flow variability, the system is effectively subjected to a quasi-stationary attack that periodically generates critical spikes. The absence of filtering means that all incoming requests, regardless of intensity, enter the system until the buffer is full. The queue is size-limited to $B = 10$, the service intensity is $\mu = 1.0$, and queue clearing occurs every 10 time steps, simulating routine checks or resets in real-world applications.

Figure 3.23 presents the averaged simulation results across 100 runs. It is evident that even in the absence of a sustained overload phase, the system experiences periodic spikes in loss probability p_{loss}, which do not exceed 0.2 but occur regularly. This indicates that the system reaches critical load at moments when λ_t approaches the upper bound of the range. By applying a normalised intensity distribution with constraints and a cap on burst request generation, artificial anomalies (such as excessively high losses) were successfully avoided.

The delay $E[T]$ responds to queue accumulation, exhibiting an inertial pattern: its increase is observed not only during overload but also after it subsides. This indicates a lag in the system's return to stability, which is typical for queues with linear processing without prioritisation or adaptive control. At the same time, queue instability $\Delta Q(t)$ shows a clear alignment with the moments of flow update, confirming the correlation between external changes and the system's internal response. Queue instability $\Delta Q(t)$, which reflects the variance in dynamics, proves to be one of the most sensitive indicators of attack. This metric responds to changes in the flow phase faster than delay and provides a potential foundation for early detection of adaptive attacks, even when the average load level is not critical. Thus, it can be regarded as a diagnostic marker of the system's fluctuation vulnerability. The minimum values

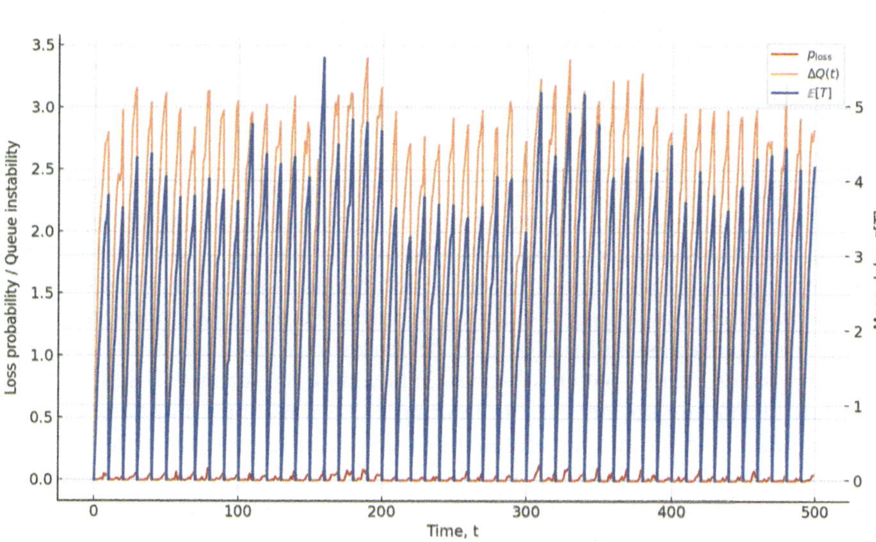

Fig. 3.23 Queue metrics under stochastic DoS attack (Phase 2, unfiltered configuration)

of delay and instability do not drop to zero even during phases of moderate load, which is explained by the residual queuing effect—the system does not have sufficient time to fully recover before the next impulse occurs. This is also due to the insufficient interval between clearing events: every 10 time steps proves inadequate to guarantee complete unloading after an impulse $\lambda_t \approx 1.5$. Such a recovery delay accumulates with each phase and manifests as increasing queue inertia, making the system progressively more vulnerable to subsequent intensity fluctuations.

The obtained results lead to the conclusion that a system without filtering is vulnerable even to stochastically oscillating attacks. Although the average load may not exceed capacity, its fluctuating nature triggers periodic phases of degradation. Such attacks are difficult to detect using standard tools, as they lack a persistent high-intensity pattern, and their effects emerge only through accumulated dynamics [292]. This operating mode indicates a critical dependence of the system on external flow fluctuations, thereby justifying the transition to a regime with filtering.

We simulate queue behaviour in phase 2 under a stochastic attack with filtering enabled. The primary aim of the experiment is to assess the extent to which limiting the input intensity can stabilise the system and reduce losses, delays, and queue instability in comparison to the unfiltered scenario discussed in the previous section. As before, the flow intensity λ_t is updated every 50 time steps; however, prior to traffic generation, a constraint is applied: all values of λ_t are filtered to a maximum $\lambda_{\max} = 1.1$. In addition, request generation follows a binomial distribution rather than a Poisson process, which limits the number of requests per time step to three and prevents burst overloads. This configuration models a protected environment with input frequency limitation at the infrastructure level, particularly at the edge tier.

Figure 3.24 shows the dynamics of queue metrics over 500 time steps. The loss probability p_{loss} remains virtually zero throughout the entire time interval. This indicates that filtering effectively protects the system from overload, even under stochastic attack conditions. Compared to the unfiltered scenario, where losses reached 0.2 in peak zones, these results demonstrate an almost complete elimination of that effect.

The average delay $E[T]$ stabilises within the range of 0.5–1.0 without exhibiting significant oscillations. This indicates that the system maintains equilibrium under dynamic conditions and compensates for fluctuating load in a controlled manner. Minor residual variations result from queue accumulation between clearing cycles; however, they do not trigger escalating failures. Particularly important is the behaviour of queue instability $\Delta Q(t)$: it decreases to a level below 1.0 and does not display the oscillatory pattern observed in the unfiltered scenario. Thus, the positive effect of filtering lies not only in limiting the average input intensity, but also in smoothing the traffic profile, eliminating burst disturbances, and reducing fluctuation amplitude [293]. This characteristic makes the filtered system significantly less sensitive to attacks with quasi-periodic or stochastically varying patterns [291]. While in the unfiltered configuration, losses periodically exceeded 0.2 and queue instability ranged between 3 and 5, under filtering these metrics approach zero. This

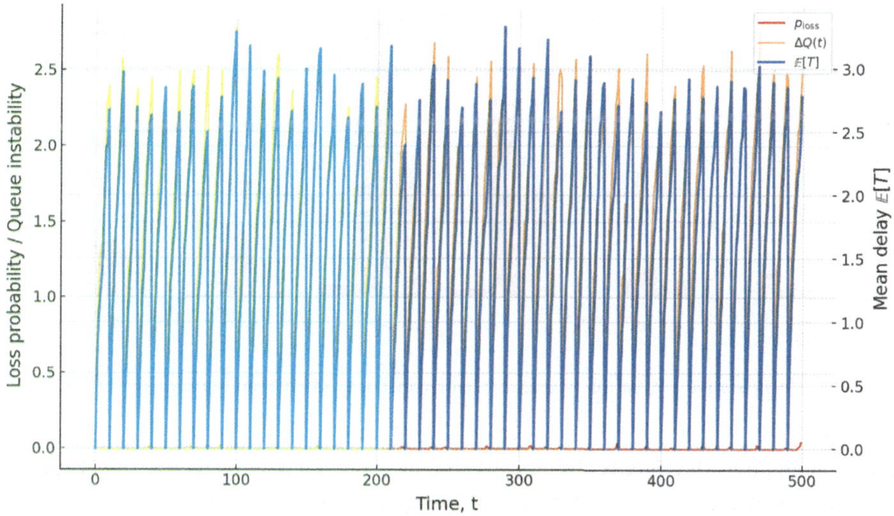

Fig. 3.24 Queue metrics under stochastic DoS attack (Phase 2, filtered configuration)

indicates increased system dependability and the ability to ensure service continuity even under background attack conditions.

At the same time, this model has a number of limitations. Firstly, it assumes a fixed filtering threshold, which makes the system potentially vulnerable to attacks adapted to that threshold. In particular, an attacking source may reduce the amplitude of the flow to remain below λ_{max}, while preserving its fluctuating structure or altering its periodicity. Secondly, filtering without feedback does not account for the actual queue state when making decisions. Thus, although simple filtering demonstrates high effectiveness in basic scenarios, its limitations create the need for an adaptive approach to flow control. The next section will examine the application of dynamic filtering, where the threshold value λ_{max} is not pre-defined but is instead determined in response to changes in internal queue metrics—specifically, queue length, instability, or average delay. This approach is intended to ensure a balance between stability and throughput in a variable environment.

The subsection concludes with a comparative analysis of system performance across six key configurations examined in the previous subsections. Three core metrics $\langle p_{loss}, E(T), \Delta Q(t)\rangle$ were selected for evaluation. All metrics were averaged over time and across simulation runs, ensuring the stability and reliability of the aggregated results.

Figure 3.25 presents the values of each metric for six configurations: phase 1 without attack, phase 2 without attack, phase 1 under attack, phase 2 under attack, phase 1 with filtering, and phase 2 with filtering. The scenarios without attacks show predictably low values across all metrics—delay does not exceed 0.30, queue instability fluctuates around 0.4–0.5, and losses are entirely absent. This indicates normal system operation without external pressure. Once attacks are introduced, the

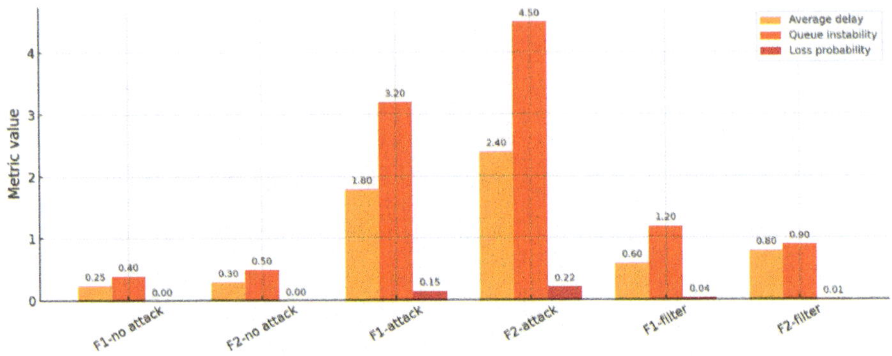

Fig. 3.25 Comparative evaluation of queue performance across six configurations

situation changes drastically: in phase 1, the average delay rises to 1.8, instability to 3.2, and losses reach 15%. In phase 2, these figures are even higher—delay reaches 2.4, instability 4.5, and the loss probability climbs to 22%. This confirms that phase 2 is the most vulnerable to attacks, as the queue accumulates more rapidly and the clearing frequency is lower.

The application of fixed filtering significantly improves all three metrics. In phase 1, the delay drops to 0.6, instability to 1.2, and losses to 4%. In phase 2, these values are even lower: delay is 0.8, instability is 0.9, and losses do not exceed 1%. This outcome demonstrates that the loss probability in phase 2 is reduced from 22 to 1%, corresponding to over 95% effectiveness [290]. Specifically, the average delay values range from 0.25 to 2.4, instability from 0.4 to 4.5, and loss probability from 0 to 0.22. Notably, filtering in phase 2 is even more effective than in phase 1: under higher initial load, the reduction of instability to below 1.0 is a significant indicator of fluctuation compensation. This reflects high efficiency not only in limiting average load but also in smoothing the burst structure of the attacked traffic. The filtering mechanism is implemented not only as a strict threshold constraint but also as a limit on the maximum number of requests per time step, modelling protective quantisation. However, despite the achieved effectiveness, fixed filtering remains insensitive to the actual state of the system. If the attacking flow alters its intensity or adapts to the pre-set threshold, the filter's effectiveness may decline. Current queue characteristics (such as length, instability, or loss rate) do not influence the filter's decisions, thus depriving the system of adaptive responsiveness. Summarising the analysis of all configurations, it can be concluded that fixed filtering provides a basic level of protection under typical stochastic attacks, but to ensure dependability under dynamic threats, the implementation of adaptive policies is required.

3.6 Summary of Simulation Study Results

Chapter 1 of the monograph outlined the general methodological framework for constructing formal Divide-and-Conquer-type models aimed at describing the behaviour of edge subsystems in 5G-IoT networks under high load, traffic fragmentation, attacks, and stochastic failures. According to this approach, the modelling system was required to enable a generalised representation of mechanisms for distribution, synchronisation, and adaptive servicing of requests. Building upon these theoretical foundations, Chap. 2 introduced the basic mathematical formalisation of servicing processes based on controlled Markovian dynamics and defined the key performance indicators: average delay $E[T]$, queue instability $\Delta Q(t)$, and loss probability p_{loss}.

The systemic adequacy of the formal models presented in Chaps. 1 and 2 of the monograph is reflected not only in their ability to reproduce individual metric dependencies, but also in their demonstration of internally consistent responses to complex external perturbations. This is confirmed by the simulation results presented in Sects. 3.1–3.5 of Appendix A, where states of synchronisation, overload, impulse attacks, stochastic losses, and variable service were analysed. In particular, the model from Sect. 1.2, which represents the processing of two classes of requests with stochastic updates and absolute priority given to attack requests, finds empirical confirmation in Sect. 3.3, where a nonlinear system response to attacks and a compensatory effect of queue clearing were recorded. This model behaviour directly corresponds to analytical formulas (1.45)–(1.47) and (1.46), which describe processing time under reactive clearing conditions. In Sect. 1.3, devoted to a cascade model with two service phases and regular clearing, an analytical dependency (formula (1.35)) for average response time was derived. Its verification is presented in Sect. 3.4, where simulation results show empirical values approaching theoretical predictions under moderate loads. At the same time, deviations under extreme parameter values highlight the justified constraints embedded in the model's clearing formalisation. The model from Sect. 1. 4, oriented towards cloud-based DC systems with a BMAP input stream, is verified in Sect. 3.5. Here, formulas (1.13) and (1.21), which describe functional availability and average recovery time under varying update intensities, are validated through simulation [35]. The results demonstrate a high level of agreement between analytical and empirical assessments. Thus, the models presented in the main part of the monograph have shown not only numerical accuracy, but also the ability to reproduce fundamental characteristics of real-world systems—such as resilience to overload, adaptability to changing traffic parameters, and preservation of structural stability. The verification studies in Sects. 3.1–3.5 support the conclusion that the proposed models are systemically adequate, which is a key prerequisite for their practical application in 5G-IoT environments.

The practical relevance of the obtained results lies primarily in the fact that the proposed formal models demonstrate the ability to operate under conditions typical of real-world edge subsystems within 5G-IoT architectures. This extends beyond traditional scenarios of increasing load or service disruptions to include situations where traffic exhibits complex correlation structures and external influences manifest as targeted attacks—for instance, through evasion requests or synchronisation delays. Simulation experiments presented in Sects. 3.3 and 3.5 confirm that the models retain the ability to isolate adversarial effects and stabilise service metrics without requiring a complete overhaul of the system's internal logic. This is particularly important given that most real-edge systems operate with limited access to centralised control and must incorporate built-in mechanisms for adaptive response to variations in load and threats. Furthermore, as shown by the results in 3.4, the models preserve their adaptability even under abrupt changes in the statistical properties of the input stream. That is, the models are not only analytically well-founded but also operationally suitable for handling traffic that violates classical assumptions such as exponential interarrival times [295]. This significantly enhances the value of the models as tools for forecasting subsystem performance in real-world conditions, where traffic is generated by heterogeneous IoT devices with diverse interaction profiles. Of particular note is the fact that the filtering mechanisms analysed in 3.5 allow for compensatory actions at the input control level without altering the core servicing architecture. This means that the DC approach can maintain scalability without sacrificing flexibility—an essential requirement for dynamic 5G-IoT environments where node configurations and load delivery routes change in real time. Overall, the practical significance of the models lies in their capacity not only to provide accurate analytical predictions of system behaviour but also to support adaptive resilience—that is, the ability to maintain control over key metrics under irregular, adversarial, or uneven load conditions.

Thus, the results obtained in Sects. 3.2–3.5 convincingly demonstrate that the objective set out in Sect. 3.1 has been fully achieved. The Divide-and-Conquer-type formal models developed and described in Chaps. 1 and 2 of the monograph have shown not only the ability to analytically describe key queuing characteristics but also to practically and reliably reproduce the dynamics of subsystems within 5G-IoT environments marked by high instability, fragmentation, and adversarial pressure. The empirical validation confirmed both the accuracy of the core metric predictions and the structural adaptability of the models to changing environmental parameters. All major experimental scenarios—from standard load conditions to delay, fluctuating input flow, and targeted attacks—were successfully reproduced without requiring modifications to the core model architecture. This attests to its systemic consistency and functional completeness. Accordingly, the DC-like approach implemented in this monograph may be regarded as a universal framework for modelling the behaviour of queuing subsystems in next-generation distributed environments. Its adequacy is supported by both theoretical constructs and empirical evidence, which allows it to be recommended for use in both research and applied engineering tasks related to ensuring service quality, resilience, and scalability in 5G-IoT infrastructures.

Postface

This monograph develops a comprehensive analytical framework for the study of Divide-and-Conquer (DC)-like queueing systems, demonstrating their applicability to modelling synchronized, multi-channel information and communication systems in the era of 5G-IoT. Through rigorous formalization, a wide range of DC-like configurations (including Split–Merge (SM), Fission–Fusion (FF), Independent Server Models (ISM), and Team Service Models (TSM)) have been modelled with progressively increasing levels of realism and mathematical sophistication.

One of the central achievements of this study lies in its successful integration of matrix-geometric methods, order statistics, and extreme value theory into the queueing analysis of distributed systems. Furthermore, the extension of input traffic models to Batch Markovian Arrival Processes (BMAP) and the inclusion of non-exponential service time distributions have significantly improved the descriptive accuracy of performance indicators such as response time, synchronization delay, and buffer overflow probabilities.

However, several inherent limitations must be acknowledged:

- Most analytical results are confined to systems with symmetrical or mildly asymmetrical service structures. In highly heterogeneous environments—where service policies, buffer capacities, or arrival processes vary significantly—the tractability of the proposed models may degrade.
- The analytical techniques rely on assumptions, such as independent subrequests, memoryless service times, or Poissonian arrivals (in base cases), which may not reflect real-world conditions characterized by burstiness, correlations, or service interruptions.
- Approximation-based approaches (e.g. H&LTIA) offer practical value but are less precise under heavy-load conditions or in systems with stringent quality-of-service constraints.
- The stochastic update models proposed for cyber-threat scenarios are limited to threshold-based or probabilistic clearing mechanisms, without addressing the full dynamics of intelligent adversarial behaviour or adaptive defence policies.

V. Kovtun, *Formal Models of Divide-and-Conquer Queuing Systems in 5G Infrastructures*, https://doi.org/10.1007/978-3-032-06187-4

These limitations, while natural in the scope of rigorous analytical modelling, offer fertile ground for future exploration.

The empirical study presented in Chap. 3 confirmed not only the practical value but also the analytical consistency of the proposed models across diverse simulation environments. By evaluating four key DC-like architectures—including baseline SM and ISM systems, a queue-cascading model with update-based recovery, and a stochastic reconfiguration model under cyberattack scenarios—the simulations validated the theoretical estimates introduced in Chaps. 1 and 2. The models successfully replicated critical behaviours observed in complex systems, such as overload resilience, loss containment, synchronization recovery, and service continuity under conditions of input anomalies and partial service degradation. Empirical metrics (including average response time, queue instability thresholds, and failure propagation) showed high congruence with analytical predictions, even under variation in input flows (Poisson and BMAP), service distributions (Erlang, H-E, Cox), and update regimes. This validation reinforces the robustness and adaptability of the DC-like approach, highlighting its scalability and fault tolerance in modelling modern 5G-IoT infrastructures subjected to real-world stressors.

The results presented in this monograph open several promising avenues for future research in the domain of queuing systems and synchronization modelling for next-generation information and communication infrastructures. One key direction involves the development of adaptive queuing models enhanced with learning mechanisms. In dynamic and heterogeneous environments such as 5G-IoT, where traffic intensity and service conditions are subject to frequent fluctuations, static queuing parameters may no longer suffice. Therefore, future systems could benefit from incorporating reinforcement learning or online estimation techniques that enable real-time adaptation of buffer sizes, service prioritization policies, and synchronization strategies based on observed traffic behaviour and system performance.

A second significant trajectory concerns the integration of DC-like queuing structures with virtualized, programmable network architectures, particularly those based on Software-Defined Networking (SDN) and Network Function Virtualization (NFV). Since these paradigms support the dynamic chaining, migration, and scaling of virtual network functions, the corresponding queuing models must account for mobility-induced delays, inter-slice interference, and context-aware service composition. Modelling DC-like systems under such virtualization constraints could bridge the gap between analytical queuing theory and real-world deployment scenarios.

Another important area of research is the formalization of queuing behaviour under strategic adversarial conditions, which arises in the context of cyber-resilience. While the current work introduces stochastic update mechanisms as an abstraction for the clearing of service queues during threat detection, more nuanced models are needed to represent intelligent adversaries. Game-theoretic formulations may offer a useful framework for capturing the interplay between attackers and defenders, allowing for the design of dynamic defence mechanisms that optimize throughput and security objectives concurrently.

Given the limitations of purely analytical techniques in capturing the full complexity of real-world traffic, it is also imperative to explore hybrid methodologies that combine theoretical results with simulation-based verification. Such hybrid platforms would allow researchers and engineers to validate closed-form approximations against empirical data, calibrate model parameters more accurately, and conduct scenario-based stress testing. In particular, digital twin environments for network systems could leverage analytical insights to guide simulation runtime decisions.

Moreover, the energy footprint of synchronized processing in edge and fog nodes introduces an additional dimension that warrants consideration. DC-like models could be extended to capture energy-delay trade-offs, especially in IoT applications where battery life is critical. Energy-aware synchronization policies that optimize subrequest allocation based on available power resources or energy harvesting patterns represent an emerging field of study with practical implications.

To complement average-case performance indicators, future research should also address worst-case performance guarantees. This can be achieved by integrating techniques from stochastic network calculus, which provide probabilistic bounds on key metrics such as delay, backlog, and packet loss. Applying these methods to DC-like architectures would offer formal assurances for time-sensitive and safety–critical applications, thereby enhancing the reliability of intelligent cyber-physical systems.

Additionally, the current modelling framework assumes a single-layer synchronization process, whereas in practice, synchronization may occur across multiple protocol layers, including application, transport, and network levels. Future work could explore hierarchical synchronization models capable of capturing the interplay between inter-layer dependencies, partial synchronization requirements, and cascading latency effects.

Finally, there is a growing need for formal verification of queuing models used in critical infrastructures. Techniques such as probabilistic model checking, temporal logic analysis, and theorem proving could be employed to ensure that synchronization mechanisms satisfy liveness, fairness, and safety properties under both normal and adversarial conditions. This line of research would not only increase the theoretical rigour of proposed models but also support their certification for use in regulated environments.

Taken together, these future directions represent a natural and necessary evolution of the analytical foundations presented in this work. They aim to align the formal modelling of divide-and-conquer queuing systems with the operational, architectural, and security demands of next-generation communication ecosystems.

By revisiting and extending the classical divide-and-conquer paradigm, this monograph contributes to the ongoing transformation of performance modelling in modern ICT systems. While several assumptions simplify real-world complexity, the proposed formalism provides a robust and extensible starting point for both theoretical advances and engineering applications. It is hoped that this work will not only stimulate further academic investigation but also assist in the practical design of intelligent, synchronized, and dependable infrastructure systems that lie at the heart of the emerging digital society.

References

1. Phyo, Y., Minh Do, C., & Ogata, K. (2021). A Divide &quasi-Toeplitz Conquer Approach to Leads-to Model Checking. In The Computer Journal (Vol. 65, Issue 6, pp. 1353–1364). Oxford University Press (OUP). https://doi.org/10.1093/comjnl/bxaa183
2. Aung, M. N., Phyo, Y., Do, C. M., & Ogata, K. (2021). A Divide and Conquer Approach to Eventual Model Checking. In Mathematics (Vol. 9, Issue 4, p. 368). MDPI AG. https://doi.org/10.3390/math9040368
3. Gao, X., Luan, H., Xia, B., Zhao, Z., Wang, J., & Meng, M. Q.-H. (2023). A divide-and-conquer control strategy with decentralized control barrier function for luggage trolley transportation by collaborative robots. In Robotica (Vol. 41, Issue 11, pp. 3333–3348). Cambridge University Press (CUP). https://doi.org/10.1017/s0263574723001005
4. Stidham, S., Jr., & Prabhu, N. U. (1974). Optimal Control of Queueing Systems. In Lecture Notes in Economics and Mathematical Systems (pp. 263–294). Springer Berlin Heidelberg. https://doi.org/10.1007/978-3-642-80838-8_13
5. Raicu, S., Costescu, D., & Popa, M. (2023). Effects of the Queue Discipline on System Performance. In AppliedMath (Vol. 3, Issue 1, pp. 37–48). MDPI AG. https://doi.org/10.3390/appliedmath3010003
6. Scala, P., Mota, M. M., & Bock, N. D. (2017). A divide and conquer approach for simulating an airport system. In International Journal of Simulation and Process Modelling (Vol. 12, Issue 6, p. 470). Inderscience Publishers. https://doi.org/10.1504/ijspm.2017.089632
7. Balsamo, S., & Marin, A. (n.d.). Queueing Networks. In Lecture Notes in Computer Science (pp. 34–82). Springer Berlin Heidelberg. https://doi.org/10.1007/978-3-540-72522-0_2
8. Smith, J. M. (2018). Introduction to Queueing Networks. In Springer Series in Operations Research and Financial Engineering. Springer International Publishing. https://doi.org/10.1007/978-3-319-78822-7
9. Bolch, G., Greiner, S., de Meer, H., & Trivedi, K. S. (2006). Queueing Networks and Markov Chains. Wiley. https://doi.org/10.1002/0471791571
10. Lin, Y., He, T., & Pang, G. (2021). Queuing Network Topology Inference Using Passive Measurements. In 2021 IFIP Networking Conference (IFIP Networking) (pp. 1–9). 2021 IFIP Networking Conference (IFIP Networking). IEEE. https://doi.org/10.23919/ifipnetworking52078.2021.9472774
11. Baron, O., Berman, O., Krass, D., & Sherzer, E. (2025). Re-direction in queueing networks with two customer types: The inter-departure analysis. In Computers &quasi-Toeplitz Operations Research (Vol. 173, p. 106867). Elsevier BV. https://doi.org/10.1016/j.cor.2024.106867

V. Kovtun, *Formal Models of Divide-and-Conquer Queuing Systems in 5G Infrastructures*, https://doi.org/10.1007/978-3-032-06187-4

12. Pathare, S. B., & Gedam, V. K. (2014). Some Estimation Approaches of Intensities for a Two Stage Open Queueing Network. In Statistics, Optimization &quasi-Toeplitz Information Computing (Vol. 2, Issue 1). International Academic Press. https://doi.org/10.19139/soic.v2i1.19
13. Thomasian, A. (2014). Analysis of Fork/Join and Related Queueing Systems. In ACM Computing Surveys (Vol. 47, Issue 2, pp. 1–71). Association for Computing Machinery (ACM). https://doi.org/10.1145/2628913
14. Hlynka, M. (2016). Queueing Theory. In Encyclopedia of Social Network Analysis and Mining (pp. 1–12). Springer New York. https://doi.org/10.1007/978-1-4614-7163-9_160-1
15. Mikki, M. A., Kangbin Yim, & Gihyun Jung. (n.d.). Popularity-independent multimedia-on-demand server model. In Proceedings 24th Annual International Computer Software and Applications Conference. COMPSAC2000 (pp. 575–580). 24th Annual International Computer Software and Applications Conference. Compsac 2000. IEEE Comput. Soc. https://doi.org/10.1109/cmpsac.2000.884783
16. Shafiq, M. T., Matthews, J., Lockley, S., & Love, P. E. D. (2018). Model server enabled management of collaborative changes in building information models. In Frontiers of Engineering Management (Vol. 0, Issue 0, p. 0). Engineering Sciences Press. https://doi.org/10.15302/j-fem-2018009
17. Chew, S. (2019). Continuous-Service M/M/1 Queuing Systems. In Applied System Innovation (Vol. 2, Issue 2, p. 16). MDPI AG. https://doi.org/10.3390/asi2020016
18. Gorodnov, V. P., Kyrylenko, V. A., & Repilo, Iu. E. (2020). The states final probabilities analytical description in queuing system with an entrance flow of requirements groups, with waiting and leaving the queue. In Radio Electronics, Computer Science, Control (Vol. 0, Issue 4, pp. 35–47). Zaporizhzhia National Technical University. https://doi.org/10.15588/1607-3274-2020-4-4
19. Srivastava, R. (2022). Service Quality Control using Queuing Theory. In Emirati Journal of Business, Economics and Social Studies (Vol. 1, Issue 1). Emirates Scholar. https://doi.org/10.54878/ejbess.169
20. Bisikalo, O. V., Kovtun, V. V., & Yukhimchuk, M. S. (2019). Modeling the security policy of the information system for critical use. In Radio Electronics, Computer Science, Control (Vol. 0, Issue 1). Zaporizhzhia National Technical University. https://doi.org/10.15588/1607-3274-2019-1-13
21. Karlis, D. (2009). A note on the exponential Poisson distribution: A nested EM algorithm. In Computational Statistics &quasi-Toeplitz Data Analysis (Vol. 53, Issue 4, pp. 894–899). Elsevier BV. https://doi.org/10.1016/j.csda.2008.10.027
22. Maya, R., Chesneau, C., Krishna, A., & Irshad, M. R. (2022). Poisson Extended Exponential Distribution with Associated INAR(1) Process and Applications. In Stats (Vol. 5, Issue 3, pp. 755–772). MDPI AG. https://doi.org/10.3390/stats5030044
23. Exponential Distribution and the Poisson Process. (2013). In Performance Modeling and Design of Computer Systems (pp. 206–224). Cambridge University Press. https://doi.org/10.1017/cbo9781139226424.016
24. Chydzinski, A., & Adamczyk, B. (2024). Response Time of Queueing Mechanisms. In Symmetry (Vol. 16, Issue 3, p. 271). MDPI AG. https://doi.org/10.3390/sym16030271
25. Kovtun, V., Izonin, I., & Gregus, M. (2022). The functional safety assessment of cyber-physical system operation process described by Markov chain. In Scientific Reports (Vol. 12, Issue 1). Springer Science and Business Media LLC. https://doi.org/10.1038/s41598-022-11193-w
26. Kovtun, V., Altameem, T., Al-Maitah, M., & Kempa, W. (2023). The Markov Concept of the Energy Efficiency Assessment of the Edge Computing Infrastructure Peripheral Server Functioning over Time. In Electronics (Vol. 12, Issue 20, p. 4320). MDPI AG. https://doi.org/10.3390/electronics12204320
27. Gao, S., Kobayashi, T., Tajiri, A., & Ota, J. (2021). Throughput analysis of conveyor systems involving multiple materials based on capability decomposition. In Computers in Industry (Vol. 132, p. 103526). Elsevier BV. https://doi.org/10.1016/j.compind.2021.103526

28. Boxma, O. J., Deng, Q., & Zwart, A. P. (2002). In Queueing Systems (Vol. 40, Issue 1, pp. 5–31). Springer Science and Business Media LLC. https://doi.org/10.1023/a:1017913826973
29. Dubovoi, V., Yukhimchuk, M., Kovtun, V., & Grochla, K. (2024). Functional Dependability of Distributed Control of Multi-Zone Objects Under Failures Conditions. In IEEE Access (Vol. 12, pp. 95736–95749). Institute of Electrical and Electronics Engineers (IEEE). https://doi.org/10.1109/access.2024.3421380
30. Bisikalo, O., Chernenko, D., Danylchuk, O., Kovtun, V., & Romanenko, V. (2020). Information Technology for TTF Optimization of an Information System for Critical Use that Operates in Aggressive Cyber-Physical Space. In 2020 IEEE International Conference on Problems of Infocommunications. Science and Technology (PIC S&quasi-ToeplitzT) (pp. 323–329). 2020 IEEE International Conference on Problems of Infocommunications. Science and Technology (PIC S&T). IEEE. https://doi.org/10.1109/picst51311.2020.9467997
31. Kovtun, V., Izonin, I., & Gregus, M. (2022). Modeling a session of subject-system interaction in a wireless communication infrastructure with a mixed resource. In H. Guo (Ed.), PLOS ONE (Vol. 17, Issue 7, p. e0271536). Public Library of Science (PLoS). https://doi.org/10.1371/journal.pone.0271536
32. Silvestrov, D., & Silvestrov, S. (2017). Asymptotic Expansions for Stationary Distributions of Nonlinearly Perturbed Semi-Markov Processes. 2. In Methodology and Computing in Applied Probability (Vol. 21, Issue 3, pp. 965–984). Springer Science and Business Media LLC. https://doi.org/10.1007/s11009-017-9607-y
33. Silvestrov, D., & Silvestrov, S. (2017). Asymptotic Expansions for Stationary Distributions of Nonlinearly Perturbed Semi-Markov Processes. 1. In Methodology and Computing in Applied Probability (Vol. 21, Issue 3, pp. 945–964). Springer Science and Business Media LLC. https://doi.org/10.1007/s11009-017-9605-0
34. Bisikalo, O. V., Kovtun, V. V., & Kovtun, O. V. (2020). Modeling of the Estimation of the Time to Failure of the Information System for Critical Use. In 2020 10th International Conference on Advanced Computer Information Technologies (ACIT) (pp. 140–143). 2020 10th International Conference on Advanced Computer Information Technologies (ACIT). IEEE. https://doi.org/10.1109/acit49673.2020.9208883
35. Kovtun, V., & Izonin, I. (2021). Study of the Operation Process of the E-Commerce Oriented Ecosystem of 5Ge Base Station, Which Supports the Functioning of Independent Virtual Network Segments. In Journal of Theoretical and Applied Electronic Commerce Research (Vol. 16, Issue 7, pp. 2883–2897). MDPI AG. https://doi.org/10.3390/jtaer16070158
36. Chiacchio, F., Oliveri, L., Khodayee, S. M., & D'Urso, D. (2022). Performance Analysis of a Repairable Production Line Using a Hybrid Dependability Queueing Model Based on Monte Carlo Simulation. In Applied Sciences (Vol. 13, Issue 1, p. 271). MDPI AG. https://doi.org/10.3390/app13010271
37. Czerniachowska, K., Wichniarek, R., & Żywicki, K. (2023). A Model for an Order-Picking Problem with a One-Directional Conveyor and Buffer. In Sustainability (Vol. 15, Issue 18, p. 13731). MDPI AG. https://doi.org/10.3390/su151813731
38. Shortle, J. F., Thompson, J. M., Gross, D., & Harris, C. M. (2018). Fundamentals of Queueing Theory. In Wiley Series in Probability and Statistics. Wiley. https://doi.org/10.1002/9781119453765
39. Bhat, U. N. (2015). An Introduction to Queueing Theory. In Statistics for Industry and Technology. Birkhäuser Boston. https://doi.org/10.1007/978-0-8176-8421-1
40. Bisikalo, O. V., Kovtun, V. V., Kovtun, O. V., & Danylchuk, O. M. (2021). Mathematical Modeling of the Availability of the Information System for Critical Use to Optimize Control of its Communication Capabilities. In International Journal of Sensors, Wireless Communications and Control (Vol. 11, Issue 5, pp. 505–517). Bentham Science Publishers Ltd. https://doi.org/10.2174/2210327910999201009163958
41. Kovtun, V., Izonin, I., & Gregus, M. (2023). Mathematical models of the information interaction process in 5G-IoT ecosystem: Different functional scenarios. In ICT Express (Vol. 9, Issue 2, pp. 264–269). Elsevier BV. https://doi.org/10.1016/j.icte.2021.11.008

42. Massey, W. A., Ekwedike, E., Hampshire, R. C., & Pender, J. J. (2022). A transient symmetry analysis for the M/M/1/k queue. In Queueing Systems (Vol. 103, Issues 1–2, pp. 1–43). Springer Science and Business Media LLC. https://doi.org/10.1007/s11134-022-09849-5
43. Resing, J., & Rietman, R. (2004). The M/M/1 queue with gated random order of service. In Statistica Neerlandica (Vol. 58, Issue 1, pp. 97–110). Wiley. https://doi.org/10.1046/j.0039-0402.2003.00112.x
44. Bisikalo, O., Danylchuk, O., Kovtun, V., Kovtun, O., Nikitenko, O., & Vysotska, V. (2022). Modeling of Operation of Information System for Critical Use in the Conditions of Influence of a Complex Certain Negative Factor. In International Journal of Control, Automation and Systems (Vol. 20, Issue 6, pp. 1904–1913). Springer Science and Business Media LLC. https://doi.org/10.1007/s12555-021-0368-6
45. Boon, M. A. A., van der Mei, R. D., & Winands, E. M. M. (2016). Heavy traffic analysis of roving server networks. In Stochastic Models (Vol. 33, Issue 2, pp. 171–209). Informa UK Limited. https://doi.org/10.1080/15326349.2016.1226142
46. Vis, P., Bekker, R., van der Mei, R. D., & Núñez-Queija, R. (2020). Heavy-traffic limits for Discriminatory Processor Sharing models with joint batch arrivals. In Operations Research Letters (Vol. 48, Issue 2, pp. 136–141). Elsevier BV. https://doi.org/10.1016/j.orl.2020.01.004
47. Zhu, J., Hu, L., Chen, Z., & Xie, H. (2022). A Queuing Model for Mixed Traffic Flows on Highways considering Fluctuations in Traffic Demand. In Y. Yang (Ed.), Journal of Advanced Transportation (Vol. 2022, pp. 1–13). Wiley. https://doi.org/10.1155/2022/4625690
48. Dong, W., Chi, X., Liu, Z., & Zhu, Y. (2014). Queuing theory based analysis for packet jitter of mixed services. In The Journal of China Universities of Posts and Telecommunications (Vol. 21, Issue 3, pp. 71–76). Elsevier BV. https://doi.org/10.1016/s1005-8885(14)60303-4
49. Arcelli, D. (2020). Exploiting Queuing Networks to Model and Assess the Performance of Self-Adaptive Software Systems: A Survey. In Procedia Computer Science (Vol. 170, pp. 498–505). Elsevier BV. https://doi.org/10.1016/j.procs.2020.03.108
50. Agresti, A. (2010). Analysis of Ordinal Categorical Data. In Wiley Series in Probability and Statistics. Wiley. https://doi.org/10.1002/9780470594001
51. Kovtun, V., Altameem, T., Al-Maitah, M., & Kempa, W. (2024). Simple statistical tests selection based parallel computating method ensures the guaranteed global extremum identification. In Journal of King Saud University - Science (Vol. 36, Issue 5, p. 103165). Scientific Scholar. https://doi.org/10.1016/j.jksus.2024.103165
52. El-Taha, M. (2023). A Review of Birth-Death and Other Markovian Discrete-Time Queues. In A. Majumder (Ed.), Advances in Operations Research (Vol. 2023, pp. 1–22). Hindawi Limited. https://doi.org/10.1155/2023/6620393
53. Pang, G., Sarantsev, A., & Suhov, Y. (2022). Birth and death processes in interactive random environments. In Queueing Systems (Vol. 102, Issues 1–2, pp. 269–307). Springer Science and Business Media LLC. https://doi.org/10.1007/s11134-022-09855-7
54. Kapodistria, S., & Palmowski, Z. (2017). Matrix geometric approach for random walks: Stability condition and equilibrium distribution. In Stochastic Models (Vol. 33, Issue 4, pp. 572–597). Informa UK Limited. https://doi.org/10.1080/15326349.2017.1359096
55. Hoshiyama, T. (2023). Cross analysis by matrix geometric method for general distribution probability in discrete time. In 2023 IEEE World AI IoT Congress (AIIoT) (pp. 0256–0258). IEEE. https://doi.org/10.1109/aiiot58121.2023.10174297
56. Vaderna, P., & Éltető, T. (2006). Transient Analysis of a Queuing System with Matrix-Geometric Methods. In Lecture Notes in Computer Science (pp. 21–33). Springer Berlin Heidelberg. https://doi.org/10.1007/11759355_5
57. Kothandaraman, D., & Kandaiyan, I. (2023). Analysis of a Heterogeneous Queuing Model with Intermittently Obtainable Servers under a Hybrid Vacation Schedule. In Symmetry (Vol. 15, Issue 7, p. 1304). MDPI AG. https://doi.org/10.3390/sym15071304
58. Peralta, O. (2022). A Markov jump process associated with the matrix-exponential distribution. In Journal of Applied Probability (Vol. 60, Issue 1, pp. 1–13). Cambridge University Press (CUP). https://doi.org/10.1017/jpr.2022.25

59. Wolfer, G., & Watanabe, S. (2024). Geometric aspects of data-processing of Markov chains. In Transactions of Mathematics and Its Applications (Vol. 8, Issue 1). Oxford University Press (OUP). https://doi.org/10.1093/imatrm/tnae001
60. Zhang, J., Fouladirad, M., & Limnios, N. (2023). A Semi-Markov Model with Geometric Renewal Processes. In Methodology and Computing in Applied Probability (Vol. 25, Issue 4). Springer Science and Business Media LLC. https://doi.org/10.1007/s11009-023-10060-z
61. Varmazyar, M., Akhavan-Tabatabaei, R., Salmasi, N., & Modarres, M. (2018). Classification and properties of acyclic discrete phase-type distributions based on geometric and shifted geometric distributions. In Journal of Industrial Engineering International (Vol. 15, Issue 4, pp. 651–665). Springer Science and Business Media LLC. https://doi.org/10.1007/s40092-018-0299-x
62. Niranjan, S. P., Latha, S. D., Vlase, S., & Scutaru, M. L. (2024). Analysis of Bulk Queueing Model with Load Balancing and Vacation. In Axioms (Vol. 14, Issue 1, p. 18). MDPI AG. https://doi.org/10.3390/axioms14010018
63. Wang, K.-H., Chuang, S.-L., & Pearn, W.-L. (2002). Maximum entropy analysis to the N policy M/G/1 queueing system with a removable server. In Applied Mathematical Modelling (Vol. 26, Issue 12, pp. 1151–1162). Elsevier BV. https://doi.org/10.1016/s0307-904x(02)00056-2
64. Boxma, O., Kella, O., & Mandjes, M. (2025). Finite customer-pool queues. In Operations Research Letters (Vol. 60, p. 107267). Elsevier BV. https://doi.org/10.1016/j.orl.2025.107267
65. Iannario, M. (2025). Ordinal Data Analysis: Statistical Perspective with Applications. In Journal of the American Statistical Association (pp. 1–4). Informa UK Limited. https://doi.org/10.1080/01621459.2024.2443255
66. Dasgupta, N., & Morrison, J. (2024). Ordinal Data Analysis. Chapman and Hall/CRC. https://doi.org/10.1201/9781003020615
67. Tikhonenko, O., & Ziolkowski, M. (2019). Queueing Models of Systems with Non-Homogeneous Customers and their Applications in Computer Science. In 2019 IEEE 15th International Scientific Conference on Informatics (pp. 000427–000432). IEEE. https://doi.org/10.1109/informatics47936.2019.9119317
68. Haviv, M., & Ritov, Y. (2001). In Queueing Systems (Vol. 38, Issue 4, pp. 495–508). Springer Science and Business Media LLC. https://doi.org/10.1023/a:1010908330518
69. Minkevičius, S. (2009). On extreme values in open queueing networks. In Mathematical and Computer Modelling (Vol. 50, Issues 7–8, pp. 1058–1066). Elsevier BV. https://doi.org/10.1016/j.mcm.2009.05.025
70. Hansen, A. (2020). The Three Extreme Value Distributions: An Introductory Review. In Frontiers in Physics (Vol. 8). Frontiers Media SA. https://doi.org/10.3389/fphy.2020.604053
71. Lee, H.-G., Arapostathis, A., & Marcus, S. I. (2008). Necessary and Sufficient Conditions for State Equivalence to a Nonlinear Discrete-Time Observer Canonical Form. In IEEE Transactions on Automatic Control (Vol. 53, Issue 11, pp. 2701–2707). Institute of Electrical and Electronics Engineers (IEEE). https://doi.org/10.1109/tac.2008.2008321
72. Kohout, J. (2023). Four-Parameter Weibull Distribution with Lower and Upper Limits Applicable in Reliability Studies and Materials Testing. In Mathematics (Vol. 11, Issue 3, p. 544). MDPI AG. https://doi.org/10.3390/math11030544
73. El-Morshedy, M., El-Dawoody, M., & El-Faheem, A. A. (2025). Symmetric and Asymmetric Expansion of the Weibull Distribution: Features and Applications to Complete, Upper Record, and Type-II Right-Censored Data. In Symmetry (Vol. 17, Issue 1, p. 131). MDPI AG. https://doi.org/10.3390/sym17010131
74. Fayomi, A., Khan, S., Tahir, M. H., Algarni, A., Jamal, F., & Abu-Shanab, R. (2024). Correction: A new extended gumbel distribution: Properties and application. In PLOS ONE (Vol. 19, Issue 8, p. e0309768). Public Library of Science (PLoS). https://doi.org/10.1371/journal.pone.0309768
75. Farahani, S. S., van den Boom, T., & De Schutter, B. (2017). On optimization of stochastic max-min-plus-scaling systems—An approximation approach. In Automatica (Vol. 83, pp. 20–27). Elsevier BV. https://doi.org/10.1016/j.automatica.2017.05.001

76. Artalejo, J. R., Economou, A., & Gómez-Corral, A. (2007). Applications of maximum queue lengths to call center management. In Computers &quasi-Toeplitz Operations Research (Vol. 34, Issue 4, pp. 983–996). Elsevier BV. https://doi.org/10.1016/j.cor.2005.05.020
77. Szavits-Nossan, J., & Grima, R. (2024). Solving stochastic gene-expression models using queueing theory: A tutorial review. In Biophysical Journal (Vol. 123, Issue 9, pp. 1034–1057). Elsevier BV. https://doi.org/10.1016/j.bpj.2024.04.004
78. Manitz, M. (2008). Queueing-model based analysis of assembly lines with finite buffers and general service times. In Computers &quasi-Toeplitz Operations Research (Vol. 35, Issue 8, pp. 2520–2536). Elsevier BV. https://doi.org/10.1016/j.cor.2006.12.016
79. Sadeghi, N., Robinson Fayek, A., & Gerami Seresht, N. (2015). Queue performance measures in construction simulation models containing subjective uncertainty. In Automation in Construction (Vol. 60, pp. 1–11). Elsevier BV. https://doi.org/10.1016/j.autcon.2015.07.023
80. Bhaskar, V., & Lallement, P. (2010). Modeling a supply chain using a network of queues. In Applied Mathematical Modelling (Vol. 34, Issue 8, pp. 2074–2088). Elsevier BV. https://doi.org/10.1016/j.apm.2009.10.019
81. Kovtun, V., Izonin, I., & Gregus, M. (2022). Model of Information System Communication in Aggressive Cyberspace: Reliability, Functional Safety, Economics. In IEEE Access (Vol. 10, pp. 31494–31502). Institute of Electrical and Electronics Engineers (IEEE). https://doi.org/10.1109/access.2022.3160837
82. Gang Wang, Zhen Zeng, & Meng Liu. (2013). Equilibrium threshold strategy in observable queueing systems in cognitive radio networks. In Proceedings of 2013 2nd International Conference on Measurement, Information and Control (ICMIC) (pp. 328–332). IEEE. https://doi.org/10.1109/mic.2013.6757976
83. Chen, S., Zhou, Z., & Bensaou, B. (2007). Stochastic RED and Its Applications. In 2007 IEEE International Conference on Communications (pp. 6362–6367). IEEE. https://doi.org/10.1109/icc.2007.1053
84. Abdel-Jaber, H., Alkhateeb, J. H., & El-Amir, M. (2022). Evaluation of the Performance for IM-RED and IGRED Algorithms using Discrete-time Queues. In 2022 14th International Conference on Computational Intelligence and Communication Networks (CICN) (pp. 23–28). 2022 14th International Conference on Computational Intelligence and Communication Networks (CICN). IEEE. https://doi.org/10.1109/cicn56167.2022.10008318
85. Bhatnagar, S., Patel, S., & Karmeshu. (2017). A stochastic approximation approach to active queue management. In Telecommunication Systems (Vol. 68, Issue 1, pp. 89–104). Springer Science and Business Media LLC. https://doi.org/10.1007/s11235-017-0377-1
86. Arita, C., & Yanagisawa, D. (2010). Exclusive Queueing Process with Discrete Time. In Journal of Statistical Physics (Vol. 141, Issue 5, pp. 829–847). Springer Science and Business Media LLC. https://doi.org/10.1007/s10955-010-0075-9
87. Bisikalo, O. V., Kovtun, V. V., & Sholota, V. V. (2019). The Information System for Critical Use Access Process Dependability Modeling. In 2019 9th International Conference on Advanced Computer Information Technologies (ACIT) (pp. 5–8). IEEE. https://doi.org/10.1109/acitt.2019.8780013
88. Kovtun, V., Izonin, I., & Gregus, M. (2022). Formalization of the metric of parameters for quality evaluation of the subject-system interaction session in the 5G-IoT ecosystem. In Alexandria Engineering Journal (Vol. 61, Issue 10, pp. 7941–7952). Elsevier BV. https://doi.org/10.1016/j.aej.2022.01.054
89. Bisikalo, O. V., Kovtun, V. V., Kovtun, O. V., & Romanenko, V. B. (2020). Research of safety and survivability models of the information system for critical use. In 2020 IEEE 11th International Conference on Dependable Systems, Services and Technologies (DESSERT). IEEE. https://doi.org/10.1109/dessert50317.2020.9125061
90. Kovtun, V., Grochla, K., Kempa, W., & Połys, K. (2023). Reliably Controlling Massive Traffic between a Sensor Network End Internet of Things Device Environment and a Hub Using Transmission Control Protocol Mechanisms. In Electronics (Vol. 12, Issue 24, p. 4920). MDPI AG. https://doi.org/10.3390/electronics12244920

91. Strielkina, A., Uzun, D., & Kharchenko, V. (2017). Modelling of healthcare IoT using the queueing theory. In 2017 9th IEEE International Conference on Intelligent Data Acquisition and Advanced Computing Systems: Technology and Applications (IDAACS). IEEE. https://doi.org/10.1109/idaacs.2017.8095207
92. Fesenko, H., Illiashenko, O., Kharchenko, V., Leichenko, K., Sachenko, A., & Scislo, L. (2024). Methods and Software Tools for Reliable Operation of Flying LiFi Networks in Destruction Conditions. In Sensors (Vol. 24, Issue 17, p. 5707). MDPI AG. https://doi.org/10.3390/s24175707
93. Kharchenko, V., Kliushnikov, I., Rucinski, A., Fesenko, H., & Illiashenko, O. (2022). UAV Fleet as a Dependable Service for Smart Cities: Model-Based Assessment and Application. In Smart Cities (Vol. 5, Issue 3, pp. 1151–1178). MDPI AG. https://doi.org/10.3390/smartcities5030058
94. Dong, F., Wu, K., & Srinivasan, V. (2017). Copula Analysis of Temporal Dependence Structure in Markov Modulated Poisson Process and Its Applications. In ACM Transactions on Modeling and Performance Evaluation of Computing Systems (Vol. 2, Issue 3, pp. 1–28). Association for Computing Machinery (ACM). https://doi.org/10.1145/3089254
95. Kovtun, V., Izonin, I., & Gregus, M. (2022). Model of functioning of the centralized wireless information ecosystem focused on multimedia streaming. In Egyptian Informatics Journal (Vol. 23, Issue 4, pp. 89–96). Elsevier BV. https://doi.org/10.1016/j.eij.2022.06.009
96. Kovtun, V. & Kovtun, O. (2024). Service-Oriented Model for Handling mMTC Subscribers' Traffic in a 5G Cluster. In Proc. 5th International Workshop on Intelligent Information Technologies & Systems of Information Security (Vol. 3675, pp. 236–246). CEUR-WS.
97. Kovtun, V., Altameem, T., Al-Maitah, M., & Kempa, W. (2023). Information Technology for Maximizing Energy Consumption for Useful Information Traffic in a Dense Wi-Fi 6/6E Ecosystem. In Electronics (Vol. 12, Issue 18, p. 3847). MDPI AG. https://doi.org/10.3390/electronics12183847
98. Kovtun, V., & Kovtun, O. (2023). Asymptotic assessment of the functional safety of information interaction in the SDH architecture at the network and transport OSI layers. In 2023 13th International Conference on Dependable Systems, Services and Technologies (DESSERT) (pp. 1–6). IEEE. https://doi.org/10.1109/dessert61349.2023.10416482
99. Breinbjerg, J., Platz, T. T., & Østerdal, L. P. (2023). Equilibrium arrivals to a last-come first-served preemptive-resume queue. In Annals of Operations Research (Vol. 336, Issue 3, pp. 1551–1572). Springer Science and Business Media LLC. https://doi.org/10.1007/s10479-023-05348-9
100. Hirbod, F., Eshghali, M., Sheikhasadi, M., Jolai, F., & Aghsami, A. (2023). A state-dependent M/M/1 queueing location-allocation model for vaccine distribution using metaheuristic algorithms. In Journal of Computational Design and Engineering (Vol. 10, Issue 4, pp. 1507–1530). Oxford University Press (OUP). https://doi.org/10.1093/jcde/qwad058
101. Ameli, M., Lebacque, J.-P., & Leclercq, L. (2021). Computational Methods for Calculating Multimodal Multiclass Traffic Network Equilibrium: Simulation Benchmark on a Large-Scale Test Case. In V. L. Knoop (Ed.), Journal of Advanced Transportation (Vol. 2021, pp. 1–17). Wiley. https://doi.org/10.1155/2021/8815653
102. Grassmann, W., & Tavakoli, J. (2022). Numerical Methods for Solving Discrete Event Systems. In CMS/CAIMS Books in Mathematics. Springer International Publishing. https://doi.org/10.1007/978-3-031-10082-6
103. Chen, Z., Xu, H., & Huo, H. (2023). Optimal queuing strategies for an M/G/1 retrial queue system with RWV and ISEV policies. In The ANZIAM Journal (Vol. 65, Issue 4, pp. 384–410). Cambridge University Press (CUP). https://doi.org/10.1017/s1446181124000014
104. Bose, S. K. (2002). An Introduction to Queueing Systems. Springer US. https://doi.org/10.1007/978-1-4615-0001-8
105. Queueing Theory 1. (2021). Wiley. https://doi.org/10.1002/9781119755432
106. Thomopoulos, N. T. (2012). Fundamentals of Queuing Systems. Springer US. https://doi.org/10.1007/978-1-4614-3713-0

107. Stidham Jr., S. (2009). Optimal Design of Queueing Systems. Chapman and Hall/CRC. https://doi.org/10.1201/9781420010008
108. Dshalalow, J. H. (2023). Advances in Queueing Theory, Methods, and Open Problems. CRC Press. https://doi.org/10.1201/9781003418283
109. Li, Q.-L., Ma, J.-Y., Fan, R.-N., & Xia, L. (2019). An Overview for Markov Decision Processes in Queues and Networks. In Communications in Computer and Information Science (pp. 44–71). Springer Singapore. https://doi.org/10.1007/978-981-15-0864-6_3
110. Mirtchev, S. T., & Ganchev, I. (2017). Generalised Pollaczek–Khinchin formula for the Polya/G/1 queue. In Electronics Letters (Vol. 53, Issue 1, pp. 27–29). Institution of Engineering and Technology (IET). https://doi.org/10.1049/el.2016.2876
111. Sakalauskas, L. L., & Minkevičius, S. (2000). On the law of the iterated logarithm in open queueing networks. In European Journal of Operational Research (Vol. 120, Issue 3, pp. 632–640). Elsevier BV. https://doi.org/10.1016/s0377-2217(99)00003-x
112. Karpelevitch, F. I., & Kreinin, A. Ya. (1996). Asymptotic analysis of queueing systems with identical service. In Journal of Applied Probability (Vol. 33, Issue 1, pp. 267–281). Cambridge University Press (CUP). https://doi.org/10.2307/3215283
113. Queueing Theory and Simulation. (2018). In Optimization Techniques and Applications with Examples (pp. 227–247). Wiley. https://doi.org/10.1002/9781119490616.ch10
114. Queueing Theory. (2021). In Operations Research (pp. 347–399). World Scientific. https://doi.org/10.1142/9789811239359_0009
115. Gallager, R. G. (2013). Stochastic Processes: Theory for Applications. Cambridge: Cambridge University Press.
116. Hu, X., Barnes, S., & Golden, B. (2017). Applying queueing theory to the study of emergency department operations: a survey and a discussion of comparable simulation studies. In International Transactions in Operational Research (Vol. 25, Issue 1, pp. 7–49). Wiley. https://doi.org/10.1111/itor.12400
117. Bailey, A. R., Alexander, A., & Shaw, G. (2019). Queuing as a Changing Shopper Experience: The Case of Grocery Shopping in Britain, 1945–1975. In Enterprise &quasi-Toeplitz Society (Vol. 20, Issue 3, pp. 652–683). Cambridge University Press (CUP). https://doi.org/10.1017/eso.2018.98
118. Haviv, M., & Ravner, L. (2021). A survey of queueing systems with strategic timing of arrivals. In Queueing Systems (Vol. 99, Issues 1–2, pp. 163–198). Springer Science and Business Media LLC. https://doi.org/10.1007/s11134-021-09717-8
119. Asanjarani, A., Nazarathy, Y., & Taylor, P. (2021). A survey of parameter and state estimation in queues. In Queueing Systems (Vol. 97, Issues 1–2, pp. 39–80). Springer Science and Business Media LLC. https://doi.org/10.1007/s11134-021-09688-w
120. Govil, M. K., & Fu, M. C. (1999). Queueing theory in manufacturing: A survey. In Journal of Manufacturing Systems (Vol. 18, Issue 3, pp. 214–240). Elsevier BV. https://doi.org/10.1016/s0278-6125(99)80033-8
121. Hsu, G.-H. (1990). A survey of queueing theory. In Annals of Operations Research (Vol. 24, Issue 1, pp. 29–43). Springer Science and Business Media LLC. https://doi.org/10.1007/bf02216814
122. Integrated Services Digital Networks. (2004). In Telecommunication System Engineering (pp. 565–602). Wiley. https://doi.org/10.1002/0471728489.ch14
123. Ronayne, J. (2023). Integrated Services Digital Network. CRC Press. https://doi.org/10.1201/9781003420408
124. Millán, G., San Juan Urrutia, E., & Vargas Guzmán, M. (2019). A Simplified Multifractal Model for Self-Similar Traffic Flows in High-Speed Computer Networks. In Computación y Sistemas (Vol. 23, Issue 4). Instituto Politecnico Nacional/Centro de Investigacion en Computacion. https://doi.org/10.13053/cys-23-4-2831
125. An, X., Qu, L., & Yan, H. (2015). A Study Based on Self-Similar Network Traffic Model. In 2015 Sixth International Conference on Intelligent Systems Design and Engineering Applications (ISDEA) (pp. 73–76). IEEE. https://doi.org/10.1109/isdea.2015.28

126. Millán, G., & Lefranc, G. (2013). A Fast Multifractal Model for Self-Similar Traffic Flows in High-Speed Computer Networks. In Procedia Computer Science (Vol. 17, pp. 420–425). Elsevier BV. https://doi.org/10.1016/j.procs.2013.05.054
127. Fonseca dos Reis, E., Li, A., & Masuda, N. (2020). Generative models of simultaneously heavy-tailed distributions of interevent times on nodes and edges. In Physical Review E (Vol. 102, Issue 5). American Physical Society (APS). https://doi.org/10.1103/physreve.102.052303
128. Nakashima, T. (2010). Queue Management for the Heavy-Tailed Traffics. In 2010 International Conference on Broadband, Wireless Computing, Communication and Applications (BWCCA 2010) (pp. 222–228). IEEE. https://doi.org/10.1109/bwcca.2010.74
129. Jelenković, P. R., & Tan, J. (2013). Characterizing Heavy-Tailed Distributions Induced by Retransmissions. In Advances in Applied Probability (Vol. 45, Issue 1, pp. 106–138). Cambridge University Press (CUP). https://doi.org/10.1239/aap/1363354105
130. Armendáriz, I., & Loulakis, M. (2011). Conditional distribution of heavy tailed random variables on large deviations of their sum. In Stochastic Processes and their Applications (Vol. 121, Issue 5, pp. 1138–1147). Elsevier BV. https://doi.org/10.1016/j.spa.2011.01.011
131. Vojta, T., Halladay, S., Skinner, S., Janušonis, S., Guggenberger, T., & Metzler, R. (2020). Reflected fractional Brownian motion in one and higher dimensions. In Physical Review E (Vol. 102, Issue 3). American Physical Society (APS). https://doi.org/10.1103/physreve.102.032108
132. Asmussen, S. (2014). Lévy Processes, Phase-Type Distributions, and Martingales. In Stochastic Models (Vol. 30, Issue 4, pp. 443–468). Informa UK Limited. https://doi.org/10.1080/15326349.2014.958424
133. Nourdin, I. (2012). Selected Aspects of Fractional Brownian Motion. Springer Milan. https://doi.org/10.1007/978-88-470-2823-4
134. Shilpa, L. S., & Srinivasan, K. (2022). Hybrid modified continuous time Markov chain model for daily streamflow generation. In Journal of Hydrology (Vol. 612, p. 128206). Elsevier BV. https://doi.org/10.1016/j.jhydrol.2022.128206
135. Hespanha, J. P. (2005). A model for stochastic hybrid systems with application to communication networks. In Nonlinear Analysis: Theory, Methods &quasi-Toeplitz Applications (Vol. 62, Issue 8, pp. 1353–1383). Elsevier BV. https://doi.org/10.1016/j.na.2005.01.112
136. Barndorff-Nielsen, O. E., Resnick, S. I., & Mikosch, T. (Eds.). (2001). Lévy Processes. Birkhäuser Boston. https://doi.org/10.1007/978-1-4612-0197-7
137. Zhang, L., & Pan, Z. (2012). Generation of IFS Fractal Images Based on Hidden Markov Model. In Lecture Notes in Computer Science (pp. 178–187). Springer Berlin Heidelberg. https://doi.org/10.1007/978-3-642-31439-1_16
138. Marwala, T., Mahola, U., & Nelwamondo, F. V. (2006). Hidden Markov Models and Gaussian Mixture Models for Bearing Fault Detection Using Fractals. In The 2006 IEEE International Joint Conference on Neural Network Proceedings (pp. 3237–3242). IEEE. https://doi.org/10.1109/ijcnn.2006.247310
139. Milne, C. (1982). Transient Behaviour of the Interrupted Poisson Process. In Journal of the Royal Statistical Society Series B: Statistical Methodology (Vol. 44, Issue 3, pp. 398–405). Oxford University Press (OUP). https://doi.org/10.1111/j.2517-6161.1982.tb01220.x
140. McNickle, D. C. (1984). Properties of the interrupted poisson approximation. In Operations Research Letters (Vol. 3, Issue 2, pp. 69–73). Elsevier BV. https://doi.org/10.1016/0167-6377(84)90044-0
141. Bhat, V. N. (1992). Switched poisson process/g/1 queue with service interruptions. In Computers &quasi-Toeplitz Operations Research (Vol. 19, Issue 8, pp. 751–756). Elsevier BV. https://doi.org/10.1016/0305-0548(92)90014-v
142. Anderson, C. W., Mole, N., & Nadarajah, S. (1997). A switching Poisson process model for high concentrations in short-range atmospheric dispersion. In Atmospheric Environment (Vol. 31, Issue 6, pp. 813–824). Elsevier BV. https://doi.org/10.1016/s1352-2310(96)00254-3
143. Choquet, R. (2018). Markov-modulated Poisson processes as a new framework for analysing capture–recapture data. In N. Yoccoz (Ed.), Methods in Ecology and Evolution (Vol. 9, Issue 4, pp. 929–935). Wiley. https://doi.org/10.1111/2041-210x.12961

144. Avanzi, B., Taylor, G., Wong, B., & Xian, A. (2021). Modelling and understanding count processes through a Markov-modulated non-homogeneous Poisson process framework. In European Journal of Operational Research (Vol. 290, Issue 1, pp. 177–195). Elsevier BV. https://doi.org/10.1016/j.ejor.2020.07.022
145. Cordeiro, J. D., & Kharoufeh, J. P. (2011). Batch Markovian Arrival Processes (<scp>BMAP</scp>). In Wiley Encyclopedia of Operations Research and Management Science. Wiley. https://doi.org/10.1002/9780470400531.eorms0096
146. Jin, F., Wu, C., & Ou, H. (2020). Compound Binomial Model with Batch Markovian Arrival Process. In W. Yu (Ed.), Mathematical Problems in Engineering (Vol. 2020, pp. 1–10). Wiley. https://doi.org/10.1155/2020/1932704
147. Kumar, N., & Gupta, U. C. (2019). Analysis of a population model with batch Markovian arrivals influenced by Markov arrival geometric catastrophes. In Communications in Statistics - Theory and Methods (Vol. 50, Issue 13, pp. 3137–3158). Informa UK Limited. https://doi.org/10.1080/03610926.2019.1682166
148. Cao, J., & Xie, W. (2018). Joint arrival process of multiple independent batch Markovian arrival processes. In Statistics &quasi-Toeplitz Probability Letters (Vol. 133, pp. 42–49). Elsevier BV. https://doi.org/10.1016/j.spl.2017.09.012
149. Chydzinski, A. (2025). Asymmetry Between Queues with Correlated Service and Correlated Arrivals. In Symmetry (Vol. 17, Issue 1, p. 106). MDPI AG. https://doi.org/10.3390/sym17010106
150. Jia, Z., Dai, W., & Hu, J.-Q. (2023). Correlated Queues with Inter-Arrival Times Depending on Service Times. In Asia-Pacific Journal of Operational Research (Vol. 41, Issue 03). World Scientific Pub Co Pte Ltd. https://doi.org/10.1142/s0217595923500227
151. Rong, Y., & Xue, Z. (2024). Study of medium and long-term free flow capacity and queue discharge rates on roads. In G. Li (Ed.), PLOS ONE (Vol. 19, Issue 2, p. e0290161). Public Library of Science (PLoS). https://doi.org/10.1371/journal.pone.0290161
152. Kovtun, V., Grochla, K., Altameem, T., & Al-Maitah, M. (2023). Evaluation of the QoS policy model of an ordinary 5G smart city cluster with predominant URLLC and eMBB traffic. In V. O. Nyangaresi (Ed.), PLOS ONE (Vol. 18, Issue 12, p. e0295252). Public Library of Science (PLoS). https://doi.org/10.1371/journal.pone.0295252
153. Kovtun, V., Grochla, K., & Połys, K. (2024). The concept of network resource control of a 5G cluster focused on the smart city's critical infrastructure needs. In Alexandria Engineering Journal (Vol. 94, pp. 248–256). Elsevier BV. https://doi.org/10.1016/j.aej.2024.03.038
154. Kovtun, V., Grochla, K., Zaitseva, E., & Levashenko, V. (2024). The concept of optimal planning of a linearly oriented segment of the 5G network. In M. Balfaqih (Ed.), PLOS ONE (Vol. 19, Issue 4, p. e0299000). Public Library of Science (PLoS). https://doi.org/10.1371/journal.pone.0299000
155. Kovtun, V., Kovtun, O., Grochla, K., & Yasniy, O. (2025). The quality of service assessment of eMBB and mMTC traffic in a clustered 5G ecosystem of a smart factory. In Egyptian Informatics Journal (Vol. 29, p. 100598). Elsevier BV. https://doi.org/10.1016/j.eij.2024.100598
156. Kovtun, V., Grochla, K., Zaitseva, E., & Levashenko, V. (2024). The concept of effective coverage radius use of the unlicensed high-frequency range in the operation of the 5G network. In Heliyon (Vol. 10, Issue 21, p. e39825). Elsevier BV. https://doi.org/10.1016/j.heliyon.2024.e39825
157. Lucantoni, D. M. (n.d.). The BMAP/G/1 queue: A tutorial. In Lecture Notes in Computer Science (pp. 330–358). Springer-Verlag. https://doi.org/10.1007/bfb0013859
158. Lucantoni, D. (1998). Further transient analysis of theBMAP/G/1 Queue. In Communications in Statistics. Stochastic Models (Vol. 14, Issues 1–2, pp. 461–478). Informa UK Limited. https://doi.org/10.1080/15326349808807482
159. Lucantoni, D. M., Choudhury, G. L., & Whitt, W. (1994). The transient BMAP/G/l queue. In Communications in Statistics. Stochastic Models (Vol. 10, Issue 1, pp. 145–182). Informa UK Limited. https://doi.org/10.1080/15326349408807291

160. Chakravarthy, S. R. (2021). Queueing Models in Services – Analytical and Simulation Approach. In Queueing Theory 2 (pp. 33–80). Wiley. https://doi.org/10.1002/9781119755234.ch2
161. Aldous, D. J. (1982). Markov chains with almost exponential hitting times. In Stochastic Processes and their Applications (Vol. 13, Issue 3, pp. 305–310). Elsevier BV. https://doi.org/10.1016/0304-4149(82)90016-3
162. Kim, B., & Kim, J. (2010). Queue size distribution in a discrete-time D-BMAP/G/1 retrial queue. In Computers &quasi-Toeplitz Operations Research (Vol. 37, Issue 7, pp. 1220–1227). Elsevier BV. https://doi.org/10.1016/j.cor.2009.04.016
163. Kim, N. K., Chaudhry, M. L., Yoon, B. K., & Kim, K. (2012). A Complete and Simple Solution to a Discrete-Time Finite-Capacity BMAP/D/c Queue. In Applied Mathematics (Vol. 03, Issue 12, pp. 2169–2173). Scientific Research Publishing, Inc. https://doi.org/10.4236/am.2012.312a297
164. Ruiz-Castro, J. E., Dawabsha, M., & Alonso, F. J. (2018). Discrete-time Markovian arrival processes to model multi-state complex systems with loss of units and an indeterminate variable number of repairpersons. In Reliability Engineering &quasi-Toeplitz System Safety (Vol. 174, pp. 114–127). Elsevier BV. https://doi.org/10.1016/j.ress.2018.02.019
165. Wu, H., & He, Q.-M. (2020). Double-sided queues with marked Markovian arrival processes and abandonment. In Stochastic Models (Vol. 37, Issue 1, pp. 23–58). Informa UK Limited. https://doi.org/10.1080/15326349.2020.1794898
166. Ruiz-Castro, J. E. (2016). Complex multi-state systems modelled through marked Markovian arrival processes. In European Journal of Operational Research (Vol. 252, Issue 3, pp. 852–865). Elsevier BV. https://doi.org/10.1016/j.ejor.2016.02.007
167. Ost, A. (2001). Quasi-Birth-and-Death Processes. In Performance of Communication Systems (pp. 51–102). Springer Berlin Heidelberg. https://doi.org/10.1007/978-3-662-04421-6_4
168. Kharoufeh, J. P. (2011). Level-Dependent Quasi-Birth-and-Death Processes. In Wiley Encyclopedia of Operations Research and Management Science. Wiley. https://doi.org/10.1002/9780470400531.eorms0460
169. Belluccini, G., López-García, M., Lythe, G., & Molina-París, C. (2022). Counting generations in birth and death processes with competing Erlang and exponential waiting times. In Scientific Reports (Vol. 12, Issue 1). Springer Science and Business Media LLC. https://doi.org/10.1038/s41598-022-14202-0
170. Absorbing Markov Chains. (2017). In Markov Chains (pp. 93–97). Wiley. https://doi.org/10.1002/9781119387596.ch6
171. Riska, A., & Smirni, E. (2002). M/G/1-Type Markov Processes: A Tutorial. In Lecture Notes in Computer Science (pp. 36–63). Springer Berlin Heidelberg. https://doi.org/10.1007/3-540-45798-4_3
172. Ramadan, M. A., El-Shazly, N. M., & Selim, B. I. (2010). A Hessenberg method for the numerical solutions to types of block Sylvester matrix equations. In Mathematical and Computer Modelling (Vol. 52, Issues 9–10, pp. 1716–1727). Elsevier BV. https://doi.org/10.1016/j.mcm.2010.06.042
173. Howell, R. W., & Schrohe, E. (2017). Unpacking Rouché's Theorem. In PRIMUS (Vol. 27, Issues 8–9, pp. 801–813). Informa UK Limited. https://doi.org/10.1080/10511970.2016.1235646
174. Thorsdottir, H., & Verloop, I. M. (2016). Markov-modulated M/G/1-type queue in heavy traffic and its application to time-sharing disciplines. In Queueing Systems (Vol. 83, Issues 1–2, pp. 29–55). Springer Science and Business Media LLC. https://doi.org/10.1007/s11134-016-9477-y
175. An Introduction to Queueing Theory and Matrix-Analytic Methods. (2005). Springer-Verlag. https://doi.org/10.1007/1-4020-3631-0
176. Chakravarthy, S. R. (2022). Introduction to Matrix-Analytic Methods in Queues 1. Wiley. https://doi.org/10.1002/9781394165421
177. He, Q.-M. (2014). Fundamentals of Matrix-Analytic Methods. Springer New York. https://doi.org/10.1007/978-1-4614-7330-5

178. Neuts, M. F. (1984). Matrix-analytic methods in queuing theory. In European Journal of Operational Research (Vol. 15, Issue 1, pp. 2–12). Elsevier BV. https://doi.org/10.1016/0377-2217(84)90034-1
179. Port, D. (2002). A Characterization of Exponential and Ordinary Generating Functions. In Journal of Combinatorial Theory, Series A (Vol. 98, Issue 2, pp. 219–234). Elsevier BV. https://doi.org/10.1006/jcta.2001.3195
180. Markov Renewal Theory. (n.d.). In An Introduction to Queueing Theory and Matrix-Analytic Methods (pp. 135–146). Springer-Verlag. https://doi.org/10.1007/1-4020-3631-0_7
181. Alsmeyer, G. (1994). On the Markov renewal theorem. In Stochastic Processes and their Applications (Vol. 50, Issue 1, pp. 37–56). Elsevier BV. https://doi.org/10.1016/0304-4149(94)90146-5
182. Semi-Markov Processes: Applications in System Reliability and Maintenance. (2015). Elsevier. https://doi.org/10.1016/c2013-0-14260-2
183. Applied Semi-Markov Processes. (2006). Springer-Verlag. https://doi.org/10.1007/0-387-29548-8
184. Koroliuk, V. S., Koroliuk, V. V., & Limnios, N. (2008). Queuing Systems with Semi-Markov Flow in Average and Diffusion Approximation Schemes. In Methodology and Computing in Applied Probability (Vol. 11, Issue 2, pp. 201–209). Springer Science and Business Media LLC. https://doi.org/10.1007/s11009-008-9081-7
185. Zhou, P., Liu, P., Wang, S., Zhang, C., Zhang, J., & Li, S. (2022). Functional state-space model for multi-channel autoregressive profiles with application in advanced manufacturing. In Journal of Manufacturing Systems (Vol. 64, pp. 356–371). Elsevier BV. https://doi.org/10.1016/j.jmsy.2022.06.014
186. Nadarajah, S. (2008). Probabilities for Queueing Systems with Embedded Markov Chains. In Stochastic Analysis and Applications (Vol. 26, Issue 3, pp. 526–536). Informa UK Limited. https://doi.org/10.1080/07362990802007079
187. Chapter 6 Renewal processes and semi-Markov processes. (1975). In Developments in Sedimentology (pp. 123–159). Elsevier. https://doi.org/10.1016/s0070-4571(08)70372-8
188. Chakravarthy, S. R., Shruti, & Rumyantsev, A. (2020). Analysis of a Queueing Model with Batch Markovian Arrival Process and General Distribution for Group Clearance. In Methodology and Computing in Applied Probability (Vol. 23, Issue 4, pp. 1551–1579). Springer Science and Business Media LLC. https://doi.org/10.1007/s11009-020-09828-4
189. Ferng, H.-W., & Chang, J.-F. (2001). In Queueing Systems (Vol. 39, Issue 2/3, pp. 109–135). Springer Science and Business Media LLC. https://doi.org/10.1023/a:1012786932415
190. Erhardsson, T. (2005). Stein's method, Markov renewal point processes, and strong memoryless times. In Lecture Notes Series, Institute for Mathematical Sciences, National University of Singapore (pp. 119–130). Co-Published with Singapore University Press. https://doi.org/10.1142/9789812567673_0007
191. Bronshtein, I. N., Semendyayev, K. A., Musiol, G., & Muehlig, H. (2004). Probability Theory and Mathematical Statistics. In Handbook of Mathematics (pp. 743–794). Springer Berlin Heidelberg. https://doi.org/10.1007/978-3-662-05382-9_16
192. Shiryayev, A. N. (Ed.). (1992). Selected Works of A. N. Kolmogorov. Springer Netherlands. https://doi.org/10.1007/978-94-011-2260-3
193. Lee, D. H. (2016). On the Two-Moment Approximation of the Discrete-Time GI/G/1 Queue with a Single Vacation. In Discrete Dynamics in Nature and Society (Vol. 2016, pp. 1–10). Wiley. https://doi.org/10.1155/2016/5345374
194. Harrison, P. G., & Bor, J. (2021). Response Time Distribution in a Tandem Pair of Queues with Batch Processing. In Journal of the ACM (Vol. 68, Issue 4, pp. 1–41). Association for Computing Machinery (ACM). https://doi.org/10.1145/3448973
195. Samanta, S. K., & Nandi, R. (2020). Queue-Length, Waiting-Time and Service Batch Size Analysis for the Discrete-Time GI/D-MSPquasi-Toeplitz^{\text {(a,b)}}/1/\infty quasi-Toeplitz Queueing System. In Methodology and Computing in Applied Probability (Vol. 23, Issue 4, pp. 1461–1488). Springer Science and Business Media LLC. https://doi.org/10.1007/s11009-020-09823-9

196. Ibrahim, R., & Whitt, W. (2011). Real-Time Delay Estimation Based on Delay History in Many-Server Service Systems with Time-Varying Arrivals. In Production and Operations Management (Vol. 20, Issue 5, pp. 654–667). SAGE Publications. https://doi.org/10.1111/j.1937-5956.2010.01196.x
197. Singh, G., Gupta, U. C., & Chaudhry, M. L. (2014). Analysis of queueing-time distributions for MAP/DN/1 queue. In International Journal of Computer Mathematics (Vol. 91, Issue 9, pp. 1911–1930). Informa UK Limited. https://doi.org/10.1080/00207160.2013.867021
198. Sørensen, R. B., Nielsen, J. J., & Popovski, P. (2020). Queuing with Deterministic Service Times and No Waiting Lines in Machine Type Communications (Version 1). arXiv. https://doi.org/10.48550/ARXIV.2009.03973
199. III.4. The M/G/2 Queueing Model. (1983). In North-Holland Mathematics Studies (pp. 319–341). Elsevier. https://doi.org/10.1016/s0304-0208(08)71877-2
200. Samanta, S. K., Chaudhry, M. L., & Pacheco, A. (2014). Analysis of B M A P/M S P/1 Queue. In Methodology and Computing in Applied Probability (Vol. 18, Issue 2, pp. 419–440). Springer Science and Business Media LLC. https://doi.org/10.1007/s11009-014-9429-0
201. Nishimura, S. (2003). AMAP/G/1 Queue with an Underlying Birth–Death Process. In Stochastic Models (Vol. 19, Issue 4, pp. 425–447). Informa UK Limited. https://doi.org/10.1081/stm-120025398
202. Machihara, F. (1999). In Queueing Systems (Vol. 33, Issue 4, pp. 277–291). Springer Science and Business Media LLC. https://doi.org/10.1023/a:1019132314203
203. Choi, B. D., Chung, Y. H., & Dudin, A. N. (2001). The BMAP/SM/1 retrial queue with controllable operation modes. In European Journal of Operational Research (Vol. 131, Issue 1, pp. 16–30). Elsevier BV. https://doi.org/10.1016/s0377-2217(99)00440-3
204. Inoue, Y., & Takine, T. (2014). The FIFO single-server queue with disasters and multiple Markovian arrival streams. In Journal of Industrial &quasi-Toeplitz Management Optimization (Vol. 10, Issue 1, pp. 57–87). American Institute of Mathematical Sciences (AIMS). https://doi.org/10.3934/jimo.2014.10.57
205. Zhou, W., Lian, Z., Xu, W., & Huang, W. (2013). A two-stage queueing network with MAP inputs and buffer sharing. In Applied Mathematical Modelling (Vol. 37, Issue 6, pp. 3736–3747). Elsevier BV. https://doi.org/10.1016/j.apm.2012.07.003
206. Vijaya Laxmi, P., & Mussa Yesuf, O. (2010). Analysis of a finite buffer general input queue with Markovian service process and accessible and non-accessible batch service. In Journal of Industrial &quasi-Toeplitz Management Optimization (Vol. 6, Issue 4, pp. 929–944). American Institute of Mathematical Sciences (AIMS). https://doi.org/10.3934/jimo.2010.6.929
207. Banerjee, A. (2016). Analysis of finite buffer queue with state dependent service and correlated customer arrivals. In Journal of the Egyptian Mathematical Society (Vol. 24, Issue 2, pp. 295–302). Egyptian Knowledge Bank. https://doi.org/10.1016/j.joems.2014.11.005
208. Zhang, M., & Hou, Z. (2011). Performance analysis of MAP/G/1 queue with working vacations and vacation interruption. In Applied Mathematical Modelling (Vol. 35, Issue 4, pp. 1551–1560). Elsevier BV. https://doi.org/10.1016/j.apm.2010.09.031
209. Banik, A. D., Gupta, U. C., & Pathak, S. S. (2006). BMAP/G/1/N queue with vacations and limited service discipline. In Applied Mathematics and Computation (Vol. 180, Issue 2, pp. 707–721). Elsevier BV. https://doi.org/10.1016/j.amc.2006.01.006
210. Zhang, J., Han, G., & Qian, Y. (2016). Queuing Theory Based Co-Channel Interference Analysis Approach for High-Density Wireless Local Area Networks. In Sensors (Vol. 16, Issue 9, p. 1348). MDPI AG. https://doi.org/10.3390/s16091348
211. Germs, R., & van Foreest, N. (2012). Analysis of finite-buffer state-dependent bulk queues. In OR Spectrum (Vol. 35, Issue 3, pp. 563–583). Springer Science and Business Media LLC. https://doi.org/10.1007/s00291-012-0282-7
212. Shin, Y. W., & Pearce, C. E. M. (1998). The BMAP/G/1 vacation queue with queue-length dependent vacation schedule. In The Journal of the Australian Mathematical Society. Series B. Applied Mathematics (Vol. 40, Issue 2, pp. 207–221). Cambridge University Press (CUP). https://doi.org/10.1017/s0334270000012479

213. Banik, A. D., & Chaudhry, M. L. (2017). Efficient Computational Analysis of Stationary Probabilities for the Queueing System BMAP/G/1/N With or Without Vacation(s). In INFORMS Journal on Computing (Vol. 29, Issue 1, pp. 140–151). Institute for Operations Research and the Management Sciences (INFORMS). https://doi.org/10.1287/ijoc.2016.0720
214. Nandi, R., Samanta, S. K., & Kim, C. (2021). Analysis of quasi-Toeplitz D quasi-Toeplitz-quasi-Toeplitz BMAP/G/1 quasi-Toeplitz queueing system under quasi-Toeplitz N quasi-Toeplitz-policy and its cost optimization. In Journal of Industrial &quasi-Toeplitz Management Optimization (Vol. 17, Issue 6, p. 3603). American Institute of Mathematical Sciences (AIMS). https://doi.org/10.3934/jimo.2020135
215. Falin, G. I. (1986). On the waiting-time process in a single-line queue with repeated calls. In Journal of Applied Probability (Vol. 23, Issue 1, pp. 185–192). Cambridge University Press (CUP). https://doi.org/10.2307/3214127
216. Falin, G. (1990). A survey of retrial queues. In Queueing Systems (Vol. 7, Issue 2, pp. 127–167). Springer Science and Business Media LLC. https://doi.org/10.1007/bf01158472
217. Kim, J., & Kim, B. (2015). A survey of retrial queueing systems. In Annals of Operations Research (Vol. 247, Issue 1, pp. 3–36). Springer Science and Business Media LLC. https://doi.org/10.1007/s10479-015-2038-7
218. Samanta, S. K., & Rashmi, K. (2025). Analytical and computational aspects of a batch arrival retrial queue with a constant retrial policy. In Quality Technology &quasi-Toeplitz Quantitative Management (pp. 1–28). Informa UK Limited. https://doi.org/10.1080/16843703.2025.2455894
219. Jailaxmi, V., & Arumuganathan, R. (2015). Analysis of a retrial queue with multiple vacations and state dependent arrivals. In RAIRO - Operations Research (Vol. 49, Issue 3, pp. 619–634). EDP Sciences. https://doi.org/10.1051/ro/2014060
220. Sun, K., Wang, J., & Zhang, Z. G. (2021). Strategic joining in a single-server retrial queue with batch service. In Journal of Industrial &quasi-Toeplitz Management Optimization (Vol. 17, Issue 6, p. 3309). American Institute of Mathematical Sciences (AIMS). https://doi.org/10.3934/jimo.2020120
221. Baruah, M., Madan, K. C., & Eldabi, T. (2014). A Batch Arrival Single Server Queue with Server Providing General Service in Two Fluctuating Modes and Reneging during Vacation and Breakdowns. In Journal of Probability and Statistics (Vol. 2014, pp. 1–12). Wiley. https://doi.org/10.1155/2014/319318
222. Masuyama, H. (2022). A new matrix-infinite-product-form solution for upper block-Hessenberg Markov chains and its quasi-algorithmic constructibility. In Advances in Applied Probability (Vol. 55, Issue 1, pp. 1–28). Cambridge University Press (CUP). https://doi.org/10.1017/apr.2022.9
223. Niranjan, S. P., Chandrasekaran, V. M., & Indhira, K. (2018). Performance characteristics of a batch service queueing system with functioning server failure and multiple vacations. In Journal of Physics: Conference Series (Vol. 1000, p. 012112). IOP Publishing. https://doi.org/10.1088/1742-6596/1000/1/012112
224. Zhao, N., Guo, Y., Lian, Z., & Wang, M. (2015). Analysis of a MAP/PH/1 queue with discretionary priority. In 2015 Winter Simulation Conference (WSC) (pp. 2872–2883). 2015 Winter Simulation Conference (WSC). IEEE. https://doi.org/10.1109/wsc.2015.7408391
225. Ye, Q. (2020). Algorithmic computation of MAP/PH/1 queue with finite system capacity and two-stage vacations. In Journal of Industrial &quasi-Toeplitz Management Optimization (Vol. 16, Issue 5, pp. 2459–2477). American Institute of Mathematical Sciences (AIMS). https://doi.org/10.3934/jimo.2019063
226. Choudhary, A., Chakravarthy, S. R., & Sharma, D. C. (2021). Analysis of MAP/PH/1 Queueing System with Degrading Service Rate and Phase Type Vacation. In Mathematics (Vol. 9, Issue 19, p. 2387). MDPI AG. https://doi.org/10.3390/math9192387
227. Bu, Q., Liu, L., & Jiang, T. (2017). A limited clearing queueing model with an orbit and non-persistent customers. In Quality Technology &quasi-Toeplitz Quantitative Management (Vol. 16, Issue 3, pp. 262–277). Informa UK Limited. https://doi.org/10.1080/16843703.2017.1396955

228. Devos, A., Walraevens, J., Phung-Duc, T., & Bruneel, H. (2020). Analysis of the queue lengths in a priority retrial queue with constant retrial policy. In Journal of Industrial &quasi-Toeplitz Management Optimization (Vol. 16, Issue 6, pp. 2813–2842). American Institute of Mathematical Sciences (AIMS). https://doi.org/10.3934/jimo.2019082
229. Dimitriou, I. (2022). The generalized join the shortest orbit queue system: stability, exact tail asymptotics and stationary approximations. In Probability in the Engineering and Informational Sciences (Vol. 37, Issue 1, pp. 154–191). Cambridge University Press (CUP). https://doi.org/10.1017/s0269964821000528
230. Baron, O., Economou, A., & Manou, A. (2016). The State-Dependent M/G/1 Queue with Orbit: Analysis and Strategic Customer Behavior. In SSRN Electronic Journal. Elsevier BV. https://doi.org/10.2139/ssrn.2887379
231. Baskar, A., & Saravanarajan, M. C. (2024). Analyzing an M/G/1 Double Orbit Retrial Queue with the Implementation of a Working Vacation Policy. In Contemporary Mathematics (pp. 3929–3952). Universal Wiser Publisher Pte. Ltd. https://doi.org/10.37256/cm.5320245276
232. Pichugin, S. B. (2023). Queuing models for subscriber links in low Earth orbit communications system. In AIP Conference Proceedings (Vol. 2549, p. 120012). XLV Academic Space Conference, Dedicated to the Memory of Academician S.P. Korolev and Other Outstanding National Scientists — Pioneers of Space Exploration. AIP Publishing. https://doi.org/10.1063/5.0117639
233. Krishnamoorthy, A., Deepak, T. G., & Joshua, V. C. (2005). AnM|G|1 Retrial Queue with Nonpersistent Customers and Orbital Search. In Stochastic Analysis and Applications (Vol. 23, Issue 5, pp. 975–997). Informa UK Limited. https://doi.org/10.1080/07362990500186753
234. Xu, J., & Liu, L. (2024). Analysis of a Two-Stage Tandem Queuing System with Priority and Clearing Service in the Second Stage. In Mathematics (Vol. 12, Issue 10, p. 1500). MDPI AG. https://doi.org/10.3390/math12101500
235. Artalejo, J. R., & Gómez-Corral, A. (2008). Retrial Queueing Systems. Springer Berlin Heidelberg. https://doi.org/10.1007/978-3-540-78725-9
236. Mitici, M., Goseling, J., van Ommeren, J.-K., de Graaf, M., & Boucherie, R. J. (2017). On a tandem queue with batch service and its applications in wireless sensor networks. In Queueing Systems (Vol. 87, Issues 1–2, pp. 81–93). Springer Science and Business Media LLC. https://doi.org/10.1007/s11134-017-9534-1
237. Park, H.-M., Kim, T.-S., & Chae, K. C. (2010). Analysis of a two-phase queueing system with a fixed-size batch policy. In European Journal of Operational Research (Vol. 206, Issue 1, pp. 118–122). Elsevier BV. https://doi.org/10.1016/j.ejor.2010.02.005
238. Chao, X., He, Q.-M., & Ross, S. (2015). Tollbooth tandem queues with infinite homogeneous servers. In Journal of Applied Probability (Vol. 52, Issue 4, pp. 941–961). Cambridge University Press (CUP). https://doi.org/10.1239/jap/1450802745
239. Saffer, Z., & Yue, W. (2014). A dual tandem queueing system with GI service time at the first queue. In Journal of Industrial &quasi-Toeplitz Management Optimization (Vol. 10, Issue 1, pp. 167–192). American Institute of Mathematical Sciences (AIMS). https://doi.org/10.3934/jimo.2014.10.167
240. Azadeh, A., Nasirian, B., Hosseinabadi Farahani, M., & Soltanpoor, M. (2015). Simulation–optimization of complex tandem queue systems with reneging and server breakdowns considering budget constraints. In Simulation (Vol. 91, Issue 10, pp. 925–941). SAGE Publications. https://doi.org/10.1177/0037549715606807
241. Palpandi, B. (2024). Performance measures for two stage tandem queue model based on fuzzy structured element. In AIP Conference Proceedings (Vol. 3023, p. 020012). 4TH Symposium on Industrial Science and Technology (SISTEC2022). AIP Publishing. https://doi.org/10.1063/5.0182908
242. Badrinath, S., & Balakrishnan, H. (2017). Control of a non-stationary tandem queue model of the airport surface. In 2017 American Control Conference (ACC) (pp. 655–661). IEEE. https://doi.org/10.23919/acc.2017.7963027

243. De Boer, P.-T. (2006). Analysis of state-independent importance-sampling measures for the two-node tandem queue. In ACM Transactions on Modeling and Computer Simulation (Vol. 16, Issue 3, pp. 225–250). Association for Computing Machinery (ACM). https://doi.org/10.1145/1147224.1147226
244. Jiang, T. (2018). Analysis of a Tollbooth Tandem Queue with Two-Class Customers and Two Heterogeneous Dedicated Servers. In Asia-Pacific Journal of Operational Research (Vol. 35, Issue 06, p. 1850043). World Scientific Pub Co Pte Lt. https://doi.org/10.1142/s0217595918500434
245. Liu, Z., Deng, W., & Chen, G. (2017). Analysis of the Optimal Resource Allocation for a Tandem Queueing System. In Q. Zhang (Ed.), Mathematical Problems in Engineering (Vol. 2017, Issue 1). Wiley. https://doi.org/10.1155/2017/5964272
246. He, Q.-M., & Chao, X. (2014). A tollbooth tandem queue with heterogeneous servers. In European Journal of Operational Research (Vol. 236, Issue 1, pp. 177–189). Elsevier BV. https://doi.org/10.1016/j.ejor.2014.01.049
247. Kamoun, F. (2012). Performance Analysis of Two Priority Queuing Systems in Tandem. In American Journal of Operations Research (Vol. 02, Issue 04, pp. 509–518). Scientific Research Publishing, Inc. https://doi.org/10.4236/ajor.2012.24060
248. Friedman, H. D. (1965). Reduction Methods for Tandem Queuing Systems. In Operations Research (Vol. 13, Issue 1, pp. 121–131). Institute for Operations Research and the Management Sciences (INFORMS). https://doi.org/10.1287/opre.13.1.121
249. Kuki, A., Sztrik, J., Bérczes, T., Tóth, Á., & Efrosinin, D. (2020). Numerical Analysis of Non-Reliable Retrial Queueing Systems with Collision and Blocking of Customers. In Journal of Mathematical Sciences (Vol. 248, Issue 1, pp. 1–13). Springer Science and Business Media LLC. https://doi.org/10.1007/s10958-020-04850-w
250. Toth, A., Sztrik, J., Kuki, A., Berczes, T., & Effosinin, D. (2019). Reliability Analysis of Finite-Source Retrial Queues with Outgoing Calls Using Simulation. In 2019 International Conference on Information and Digital Technologies (IDT) (pp. 504–511). IEEE. https://doi.org/10.1109/dt.2019.8813419
251. Kuki, A., Bérczes, T., Sztrik, J., & Kvach, A. (2019). Numerical Analysis of Retrial Queueing Systems with Conflict of Customers and an Unreliable Server. In Journal of Mathematical Sciences (Vol. 237, Issue 5, pp. 673–683). Springer Science and Business Media LLC. https://doi.org/10.1007/s10958-019-04193-1
252. Ausín, M. C. (2014). Queues in Reliability. In Wiley StatsRef: Statistics Reference Online. Wiley. https://doi.org/10.1002/9781118445112.stat04257
253. Bhagat, A. (2020). Unreliable priority retrial queues with double orbits and discouraged customers. In AIP Conference Proceedings (Vol. 2214, p. 020014). Advancements in Mathematics and Its Emerging Areas. AIP Publishing. https://doi.org/10.1063/5.0003372
254. Nishizaki, K. (2012). Secure Implementation in Queueing Problems. In Theoretical Economics Letters (Vol. 02, Issue 05, pp. 561–565). Scientific Research Publishing, Inc. https://doi.org/10.4236/tel.2012.25103
255. Bailis, P., & Kingsbury, K. (2014). The Network is Reliable. In Queue (Vol. 12, Issue 7, pp. 20–32). Association for Computing Machinery (ACM). https://doi.org/10.1145/2639988.2655736
256. Johnstone, M. (2009). A dynamic architecture for increased passenger queue model fidelity. In Proceedings of the 2009 Winter Simulation Conference (WSC 2009) (pp. 3129–3139). IEEE. https://doi.org/10.1109/wsc.2009.5429278
257. Altman, E. (n.d.). Applications of Dynamic Games in Queues. In Annals of the International Society of Dynamic Games (pp. 309–342). Birkhäuser Boston. https://doi.org/10.1007/0-8176-4429-6_19
258. Limlawan, V., & Anussornnitisarn, P. (2020). Increase the System Utilization by Adaptive Queue Management System with Time Restricted Reservation. In ICAAI 2020: 2020 The 4th International Conference on Advances in Artificial Intelligence (pp. 53–58). ACM. https://doi.org/10.1145/3441417.3441424

259. Marek, D., Szyguła, J., Domański, A., Domańska, J., Filus, K., & Szczygieł, M. (2022). Adaptive Hurst-Sensitive Active Queue Management. In Entropy (Vol. 24, Issue 3, p. 418). MDPI AG. https://doi.org/10.3390/e24030418
260. Aissani, A., Braham, K. A., & Taleb, S. (2020). On a Queueing System with Differentiated Vacations and General Service Time. In 2020 International Conference on Advanced Computing and Applications (ACOMP) (pp. 7–14). IEEE. https://doi.org/10.1109/acomp50827.2020.00009
261. Tian, N., & Zhang, Z. G. (2006). Vacation Queueing Models Theory and Applications. In International Series in Operations Research &quasi-Toeplitz Management Science. Springer US. https://doi.org/10.1007/978-0-387-33723-4
262. Sundararaman, M., Narasimhan, D., & Rajadurai, P. (2024). Performance Analysis of Two Different Types of Waiting Queues with Working Vacations. In A. Di Crescenzo (Ed.), Journal of Mathematics (Vol. 2024, pp. 1–19). Hindawi Limited. https://doi.org/10.1155/2024/5829171
263. Goswami, V., & Panda, G. (2022). Synchronized abandonment in discrete-time renewal input queues with vacations. In Journal of Industrial and Management Optimization (Vol. 18, Issue 6, p. 4373). American Institute of Mathematical Sciences (AIMS). https://doi.org/10.3934/jimo.2021163
264. Gray, W. J., Wang, P. P., & Scott, M. (2000). A vacation queueing model with service breakdowns. In Applied Mathematical Modelling (Vol. 24, Issues 5–6, pp. 391–400). Elsevier BV. https://doi.org/10.1016/s0307-904x(99)00048-7
265. Yue, D., Yue, W., & Zhao, G. (2015). Analysis of an M/M/1 queue with vacations and impatience timers which depend on the server's states. In Journal of Industrial and Management Optimization (Vol. 12, Issue 2, pp. 653–666). American Institute of Mathematical Sciences (AIMS). https://doi.org/10.3934/jimo.2016.12.653
266. Chakravarthy, S. R. (2009). Analysis of a multi-server queue with Markovian arrivals and synchronous phase type vacations. In Asia-Pacific Journal of Operational Research (Vol. 26, Issue 01, pp. 85–113). World Scientific Pub Co Pte Ltd. https://doi.org/10.1142/s0217595909002134
267. Li, J., & Liu, L. (2017). On the GI/M/1 Queue with Vacations and Multiple Service Phases. In J.-C. Cortés (Ed.), Mathematical Problems in Engineering (Vol. 2017, Issue 1). Wiley. https://doi.org/10.1155/2017/3246934
268. Carlucci, G., De Cicco, L., & Mascolo, S. (2018). Controlling queuing delays for real-time communication. In ACM SIGCOMM Computer Communication Review (Vol. 46, Issue 3, pp. 1–7). Association for Computing Machinery (ACM). https://doi.org/10.1145/3243157.3243158
269. Abbas, G., Halim, Z., & Abbas, Z. H. (2016). Fairness-Driven Queue Management: A Survey and Taxonomy. In IEEE Communications Surveys &quasi-Toeplitz Tutorials (Vol. 18, Issue 1, pp. 324–367). Institute of Electrical and Electronics Engineers (IEEE). https://doi.org/10.1109/comst.2015.2463121
270. Ngwenya, D. W., Hlophe, M. C., & Maharaj, B. T. (2025). TCP Congestion Control Algorithm Using Queueing Theory-Based Optimality Equation. In Electronics (Vol. 14, Issue 2, p. 263). MDPI AG. https://doi.org/10.3390/electronics14020263
271. Saif, S., Das, P., Biswas, S., Khan, S., Haq, M. A., & Kovtun, V. (2024). A secure data transmission framework for IoT enabled healthcare. In Heliyon (Vol. 10, Issue 16, p. e36269). Elsevier BV. https://doi.org/10.1016/j.heliyon.2024.e36269
272. Kovtun, V., Grochla, K., Aldosary, S., & Al-Maitah, M. (2024). Analysis of direct traffic at the transport protocol level in the WiMax-1/2 cluster oriented to offload the smart city's wireless ecosystem. In Royal Society Open Science (Vol. 11, Issue 7). The Royal Society. https://doi.org/10.1098/rsos.240206
273. Kovtun, V., Kovtun, O., Grochla, K., & Połys, K. (2024). Assessment of Communication Resource Allocation by the Transmission Control Protocol for the Target Virtual Connection under Competitive Conditions. In Electronics (Vol. 13, Issue 7, p. 1180). MDPI AG. https://doi.org/10.3390/electronics13071180

274. Kovtun, V., Grochla, K., & Połys, K. (2023). Investigation of the Information Interaction of the Sensor Network End IoT Device and the Hub at the Transport Protocol Level. In Electronics (Vol. 12, Issue 22, p. 4662). MDPI AG. https://doi.org/10.3390/electronics12224662
275. Raynaud, A., Serru, T., & Nguyen, N. (2024). Attack Scenarios Generation Algorithm Based on Discrete Event System Formalism. In ACM SIGAda Ada Letters (Vol. 43, Issue 2, pp. 100–104). Association for Computing Machinery (ACM). https://doi.org/10.1145/3672359.3672376
276. Gu, C., Jiang, L., Tan, R., Li, M., & Huang, J. (2021). Attack-aware Synchronization-free Data Timestamping in LoRaWAN. In ACM Transactions on Sensor Networks (Vol. 18, Issue 1, pp. 1–31). Association for Computing Machinery (ACM). https://doi.org/10.1145/3474368
277. Pereira, Á. L. de S., Brandão, C. C. S., Ginoris, Y. P., & Alves, C. P. P. (2023). Upflow Direct Filtration and Downflow Direct Filtration: A Comparison with Emphasis on the Removal of Aerobic Spore-Forming Bacteria and Fluorescent Microspheres as Surrogates for Cryptosporidium Oocyst Removal. In Water (Vol. 15, Issue 22, p. 4012). MDPI AG. https://doi.org/10.3390/w15224012
278. Imamverdiyev, Y., & Nabiyev, B. (2016). Queuing model for information security monitoring systems. In Problems of Information Technology (Vol. 07, Issue 1, pp. 28–32). Institute of Information Technology, Ministry of Science and Education of the Republic of Azerbaijan. https://doi.org/10.25045/jpit.v07.i1.04
279. Wided, A., Assia, B., & Fatima, B. (2023). Traffic Management system and Traffic Light Control in Smart City to Reduce Traffic Congestion. In International Journal of Automation and Smart Technology (Vol. 13, Issue 1, pp. 2464–2464). Chinese Institute of Automation Engineers. https://doi.org/10.5875/ausmt.v13i1.2464
280. Levin, M. W. (2022). A general maximum-stability dispatch policy for shared autonomous vehicle dispatch with an analytical characterization of the maximum throughput. In Transportation Research Part B: Methodological (Vol. 163, pp. 258–280). Elsevier BV. https://doi.org/10.1016/j.trb.2022.07.003
281. Zeisberger, S. (2022). Do people care about loss probabilities? In Journal of Risk and Uncertainty (Vol. 65, Issue 2, pp. 185–213). Springer Science and Business Media LLC. https://doi.org/10.1007/s11166-022-09391-y
282. Yu, F., Tan, B., He, T., He, S., Huang, Y., Cai, S., & Lin, H. (2025). A Wide-Range Adjustable Conservative Memristive Hyperchaotic System with Transient Quasi-Periodic Characteristics and Encryption Application. In Mathematics (Vol. 13, Issue 5, p. 726). MDPI AG. https://doi.org/10.3390/math13050726
283. Valdovinos, I. A., Pérez-Díaz, J. A., Choo, K.-K. R., & Botero, J. F. (2021). Emerging DDoS attack detection and mitigation strategies in software-defined networks: Taxonomy, challenges and future directions. In Journal of Network and Computer Applications (Vol. 187, p. 103093). Elsevier BV. https://doi.org/10.1016/j.jnca.2021.103093
284. Celikoglu, H. B. (2014). Dynamic Classification of Traffic Flow Patterns Simulated by a Switching Multimode Discrete Cell Transmission Model. In IEEE Transactions on Intelligent Transportation Systems (Vol. 15, Issue 6, pp. 2539–2550). Institute of Electrical and Electronics Engineers (IEEE). https://doi.org/10.1109/tits.2014.2317850
285. Kovtun, V., & Kovtun, O. (2024). The Concept of Efficient Utilization of the Uplink Frequency Resource of a Smart Factory 5G Cluster by IIoT Devices. In Proc. 8th International Conference on Computational Linguistics and Intelligent Systems (Vol. 3664, pp. 273–283). CEUR-WS.
286. Kovtun, V., & Kovtun, O. (2024) Service-Oriented Model for Handling mMTC Subscribers' Traffic in a 5G Cluster. In Proc. 5th International Workshop on Intelligent Information Technologies & Systems of Information Security (Vol. 3675, pp. 236–246). CEUR-WS.

The manufacturer's authorised representative in the EU is Springer
Nature Customer Service Centre GmbH, Europaplatz 3, 69115 Heidelberg,
Germany. If you have any concerns regarding our products, please
contact ProductSafety@springernature.com

Printed and bound by CPI Group (UK) Ltd, Croydon, CR0 4YY
17/09/2026
02218438-0009